Significant Figures

By the same author

Concepts of Modern Mathematics
Does God Play Dice?
Fearful Symmetry (with Martin Golubitsky)
Game, Set, and Math
Another Fine Math You've Got Me Into
Nature's Numbers
From Here to Infinity
The Magical Maze
Life's Other Secret
Flatterland
What Shape is a Snowflake? (revised edition: *The Beauty of Numbers in Nature*)
The Annotated Flatland
Math Hysteria
The Mayor of Uglyville's Dilemma
How to Cut a Cake
Letters to a Young Mathematician
Taming the Infinite (alternative title: *The Story Of Mathematics*)
Why Beauty is Truth
Cows in the Maze
Professor Stewart's Cabinet of Mathematical Curiosities
Mathematics of Life
Professor Stewart's Hoard of Mathematical Treasures
Seventeen Equations that Changed the World (alternative title: *In Pursuit of the Unknown*)
The Great Mathematical Problems (alternative title: *Visions of Infinity*)
Symmetry: a Very Short Introduction
Jack of All Trades (science fiction eBook)
Professor Stewart's Casebook of Mathematical Mysteries
Professor Stewart's Incredible Numbers
Calculating the Cosmos
Infinity: a Very Short Introduction

With Jack Cohen
The Collapse Of Chaos
Evolving the Alien (alternative title: *What Does A Martian Look Like?*)
Figments of Reality
Wheelers (science fiction)
Heaven (science fiction)

Science Of Discworld series (with Terry Pratchett and Jack Cohen)
The Science of Discworld
The Science of Discworld II: The Globe
The Science of Discworld III: Darwin's Watch
The Science of Discworld IV: Judgement Day

With Tim Poston
The Living Labyrinth (science fiction)
Rock Star (science fiction)

iPad app
Incredible Numbers

Significant Figures

Lives and Works of Trailblazing Mathematicians

Ian Stewart

P

PROFILE BOOKS

First published in Great Britain in 2017 by
PROFILE BOOKS LTD
3 Holford Yard
Bevin Way
London
WC1X 9HD
www.profilebooks.com

ISBN 978 178125 429 5
eISBN 978 178283 149 5
Export ISBN 978 178125 899 6
Typeset in Stone Serif by Data Standards Ltd, Frome, Somerset
Printed and bound in Britain by Clays, Bungay, Suffolk

The paper this book is printed on is certified by the © 1996 Forest Stewardship Council A. C. (FSC). It is ancient-forest friendly. The printer holds FSC chain of custody SGS-COC-2061.

Contents

To John Davey, editor and friend
(19 April 1945–21 April 2017)

Introduction

ALL BRANCHES OF SCIENCE can trace their origins far back into the mists of history, but in most subjects the history is qualified by 'we now know this was wrong' or 'this was along the right lines, but today's view is different'. For example, the Greek philosopher Aristotle thought that a trotting horse can never be entirely off the ground, which Eadweard Muybridge disproved in 1878 using a line of cameras linked to tripwires. Aristotle's theories of motion were completely overturned by Galileo Galilei and Isaac Newton, and his theories of the mind bear no useful relation to modern neuroscience and psychology.

Mathematics is different. It endures. When the ancient Babylonians worked out how to solve quadratic equations – probably around 2000 BC, although the earliest tangible evidence dates from 1500 BC – their result never became obsolete. It was correct, and they knew why. It's still correct today. We express the result symbolically, but the reasoning is identical. There's an unbroken line of mathematical thought that goes all the way back from tomorrow to Babylon. When Archimedes worked out the volume of a sphere, he didn't use algebraic symbols, and he didn't think of a specific *number* π as we now do. He expressed the result geometrically, in terms of proportions, as was Greek practice then. Nevertheless, his answer is instantly recognisable as being equivalent to today's $^4/_3\pi r^3$.

To be sure, a few ancient discoveries outside mathematics have been similarly long-lived. Archimedes's Principle that an object displaces its own weight of liquid is one, and his law of the lever is another. Some parts of Greek physics and engineering live on too. But in those subjects, longevity is the exception, whereas in mathematics it's closer to the rule. Euclid's *Elements*, laying out a logical basis for geometry, still repays close examination. Its theorems remain true, and many remain useful. In mathematics, we move on, but we don't discard our history.

Before you all start to think that mathematics is burying its head in the past, I need to point out two things. One is that the perceived importance of a method or a theorem can change. Entire areas of mathematics have gone out of fashion, or become obsolete as the frontiers shifted or new techniques took over. But they're still *true*, and from time to time an obsolete area has undergone a revival, usually because of a newly discovered connection with another area, a new application, or a breakthrough in methodology. The second is that as mathematicians have developed their subject, they've not only moved on; they've also devised a gigantic amount of new, important beautiful, and useful mathematics.

That said, the basic point remains unchallenged: once a mathematical theorem has been correctly proved, it becomes something that we can build on – forever. Even though our concept of proof has tightened up considerably since Euclid's day, to get rid of unstated assumptions, we can fill in what we now see as gaps, and the results still stand.

Significant Figures investigates the almost mystical process that brings new mathematics into being. Mathematics doesn't arise in a vacuum: it's created by *people*. Among them are some with astonishing originality and clarity of mind, the people we associate with great breakthroughs – the pioneers, the trailblazers, the significant figures. Historians rightly explain that the work of the greats depended on a vast supporting cast, contributing tiny bits and pieces to the overall puzzle. Important or fruitful questions can be stated by relative unknowns; major ideas can be dimly perceived by people who lack the technical ability to turn them into powerful new methods and viewpoints. Newton remarked that he 'stood on the shoulders of giants'. He was to some extent being sarcastic; several of those giants (notably Robert Hooke) were complaining that Newton was not so much standing on their shoulders as treading on their toes, by not giving them fair credit, or by taking the credit in public despite citing their contributions in his writings. However, Newton spoke truly: his great syntheses of motion, gravity, and light depended on a huge number of insights from his intellectual predecessors. Nor were they exclusively giants. Ordinary people played a significant part too.

Nevertheless, the giants stand out, leading the way while the rest of us follow. Through the lives and works of a selection of significant figures, we can gain insight into how new mathematics is created, who created it, and how they lived. I think of them not just as pioneers who showed the rest of us the way, but as trailblazers who hacked traversable paths through the tangled undergrowth of the sprawling jungle of mathematical thought. They spent much of their time struggling through thorn bushes and swamps, but from time to time they came across a Lost City of the Elephants or an El Dorado, uncovering precious jewels hidden among the undergrowth. They penetrated regions of thought previously unknown to humankind.

Indeed, they *created* those regions. The mathematical jungle isn't like the Amazon Rainforest or the African Congo. The mathematical trailblazer isn't a David Livingstone, hacking a route along the Zambezi or hunting for the source of the Nile. Livingstone was 'discovering' things that were *already there*. Indeed, the local inhabitants knew they were there. But in those days, Europeans interpreted 'discovery' as 'Europeans bringing things to the attention of other Europeans.' Mathematical trailblazers don't merely explore a pre-existing jungle. There's a sense in which they create the jungle as they proceed; as if new plants are springing to life in their footsteps, rapidly becoming saplings, then mighty trees. However, it *feels* as if there's a pre-existing jungle, because you don't get to choose which plants spring to life. You choose where to tread, but you can't decide to 'discover' a clump of mahogany trees if what actually turns up there is a mangrove swamp.

This, I think, is the source of the still popular Platonist view of mathematical ideas: that mathematical truths 'really' exist, but they do so in an ideal form in some sort of parallel reality, which has always existed and always will. In this view, when we prove a new theorem we just find out what has been there all along. I don't think that Platonism makes literal sense, but it accurately describes the process of mathematical research. You don't get to choose: all you can do is shake the bushes and see if anything drops out. In *What is Mathematics, Really?* Reuben Hersh offers a more realistic view of mathematics: it's a shared human mental construct. In this respect it's much like money. Money isn't 'really' lumps of metal or pieces of paper or numbers in a computer; it's a shared set of conventions about how we exchange lumps of metal, pieces of paper, and numbers in a computer, for each other or for goods.

Hersh outraged some mathematicians, who zoomed in on 'human construct' and complained that mathematics is by no means arbitrary. Social relativism doesn't hack it. This is true, but Hersh explained perfectly clearly that mathematics isn't *any* human construct. We choose to tackle Fermat's Last Theorem, but we don't get to choose whether it's true or false. The human construct that is mathematics is subject to a stringent system of logical constraints, and something gets added to the construct only if it respects those constraints. Potentially, the constraints allow us to distinguish truth from falsity, but we don't find out which of those applies by declaiming loudly that only one of them is possible. The big question is: which one? I've lost count of the number of times someone has attacked some controversial piece of mathematics that they dislike by pointing out that mathematics is a tautology: everything new is a logical consequence of things we already know. Yes, it is. The new is implicit in the old. But the hard work comes when you want to make it explicit. Ask Andrew Wiles; it's no use telling him that the status of Fermat's Last Theorem was always predetermined by the logical structure of mathematics. He spent seven years finding out what its predetermined status *is*. Until you do that, being predetermined is as useful as asking someone the way to the British Library and being told that it's in Britain.

Significant Figures isn't an organised history of the whole of mathematics, but I've tried to present the mathematical topics that arise in a coherent manner, so that the concepts build up systematically as the book proceeds. On the whole, this requires presenting everything in roughly chronological order. Chronological order by topic would be unreadable, because we'd be perpetually hopping from one mathematician to another, so I've ordered the chapters by birth date and provided occasional cross-references.[1]

My significant figures are 25 in number, ancient and modern, male and female, eastern and western. Their personal histories begin in ancient Greece, with the great geometer and engineer Archimedes, whose achievements ranged from approximating π and calculating the area and volume of a sphere, to the Archimedean screw for raising water and a crane-like machine for destroying enemy ships. Next come three representatives of the far east, where the main mathematical action of the Middle Ages took place: the Chinese scholar Liu Hui, the Persian mathematician Muhammad ibn Musa Al-Khwarizmi, whose works gave us the words 'algorithm' and 'algebra', and the Indian Madhava of Sangamagrama, who pioneered infinite series for trigonometric functions, rediscovered in the west by Newton a millennium later.

The main action in mathematics returned to Europe during the Italian Renaissance, where we encounter Girolamo Cardano, one of the biggest rogues ever to grace the mathematical pantheon. A gambler and brawler, Cardano also wrote one of the most important algebra texts ever printed, practised medicine, and led a life straight out of the tabloid press. He cast horoscopes, too. In contrast, Pierre de Fermat, famous for his Last Theorem, was a lawyer with a passion for mathematics that often led him to neglect his legal work. He turned number theory into a recognised branch of mathematics, but also contributed to optics and developed some precursors to calculus. That subject was brought to fruition by Newton, whose masterwork is his *Philosophiae Naturalis Principia Mathematica* (Mathematical Principles of Natural Philosophy), usually abbreviated to *Principia*. In it, he stated his laws of motion and gravity, and applied them to the motion of the solar system. Newton marks a tipping point in mathematical physics, turning it into an organised mathematical study of what he called the 'System of the World'.

For a century after Newton, the focus of mathematics shifted to continental Europe and Russia. Leonhard Euler, the most prolific mathematician in history, turned out important papers at a journalistic rate, while systematising many areas of mathematics in a series of elegant, clearly written textbooks. No field of mathematics evaded his scrutiny. Euler even anticipated some of the ideas of Joseph Fourier, whose investigation of the transmission of heat led to one of the most important techniques in the modern engineer's handbook: Fourier analysis, which represents a periodic waveform in terms of the basic trigonometric functions 'sine' and 'cosine'. Fourier was also the first to understand that the atmosphere plays an important role in the Earth's heat balance.

Mathematics enters the modern era with the peerless researches of Carl Friedrich Gauss, a strong contender for the greatest mathematician of all time. Gauss began in number theory, sealed his reputation in celestial mechanics by predicting the reappearance of the newly discovered asteroid Ceres, and made major advances regarding complex numbers, least-squares data fitting, and non-Euclidean geometry, though he published nothing on the latter because he feared it was too far ahead of its time and would attract ridicule. Nikolai Ivanovich Lobachevsky was less diffident, and published extensively on an alternative geometry to that of Euclid, now called hyperbolic geometry. He and Janós Bolyai are now recognised as the rightful founders of non-Euclidean geometry, which can be interpreted as the natural geometry of a surface with constant curvature. Gauss was basically right to believe that the idea was ahead of its time, however, and neither Lobachevsky nor Bolyai was appreciated during his lifetime. We round off this era with the tragic story of the revolutionary Évariste Galois, killed at the age of twenty in a duel over a young woman. He made major advances in algebra, leading to today's characterisation of the vital concept of symmetry in terms of transformation groups.

A new theme now enters the story, a trail blazed by the first female mathematician we encounter. Namely, the mathematics of computation. Augusta Ada King, Countess of Lovelace, acted as assistant to Charles Babbage, a single-minded individual who understood the potential power of calculating machines. He envisaged the Analytical Engine, a programmable computer made of ratchets and cogwheels, now the central gimmick of steampunk science fiction. Ada is widely credited with being the first computer programmer, although that claim is controversial. The computer theme continues with George Boole, whose *Laws of Thought* laid down a fundamental mathematical formalism for the digital logic of today's computers.

As mathematics becomes more diverse, so does our tale, hacking its way into new regions of the ever-growing jungle. Bernhard

Riemann was brilliant at uncovering simple, general ideas behind apparently complex concepts. His contributions include the foundations of geometry, especially the curved 'manifolds' upon which Albert Einstein's revolutionary theory of gravitation, General Relativity, depends. But he also made huge steps in the theory of prime numbers by relating number theory to complex analysis through his 'zeta function'. The Riemann Hypothesis, about the zeros of this function, is one of the greatest and most important unsolved problems in the whole of mathematics, with a million-dollar prize for its solution.

Next comes Georg Cantor, who changed the way mathematicians think about the foundations of their subject by introducing set theory, and defined infinite analogues of the counting numbers 1, 2, 3, ... , leading to the discovery that some infinities are bigger than others – in a rigorous, meaningful, and useful sense. Like many innovators, Cantor was misunderstood and ridiculed during his lifetime.

Our second woman mathematician now appears on the scene, the prodigiously talented Sofia Kovalevskaia. Her life was rather complicated, tied up with Russian revolutionary politics and the obstacles that male-dominated society placed in the path of brilliant female intellectuals. It's amazing that she accomplished anything in mathematics at all. In fact, she made remarkable discoveries in the solution of partial differential equations, the motion of a rigid body, the structure of the rings of Saturn, and the refraction of light by a crystal.

The story now gathers pace. Around the turn of the nineteenth century, one of the world's leading mathematicians was the Frenchman Henri Poincaré. An apparent eccentric, he was actually extremely shrewd. He recognised the importance of the nascent area of topology – 'rubber-sheet geometry' in which shapes can be distorted continuously – and extended it from two dimensions to three and beyond. He applied it to differential equations, studying the three-body problem for Newtonian gravitation. This led him to discover the possibility of deterministic chaos, apparently random behaviour in a non-random system. He also came close to discovering Special Relativity before Einstein did.

As a German counterpart to Poincaré we have David Hilbert,

whose career divides into five distinct periods. First, he took up a line of thought that originated with Boole, about 'invariants' – algebraic expressions that remain the same despite changes in coordinates. He then developed a systematic treatment of core areas of number theory. After that, he revisited Euclid's axioms for geometry, found them wanting, and added extra ones to plug the logical gaps. Next, he moved into mathematical logic and foundations, initiating a programme to prove that mathematics can be placed on an axiomatic basis, and that this is both consistent (no logical deduction can lead to a contradiction) and complete (every statement can either be proved or disproved). Finally, he turned to mathematical physics, coming close to beating Einstein to General Relativity, and introducing the notion of a Hilbert space, central to quantum mechanics.

Emmy Noether is our third and final female mathematician, who lived at a time when the participation of women in academic matters was still frowned upon by most of the incumbent males. She began, like Hilbert, in invariant theory, and later worked with him as a colleague. Hilbert made strenuous attempts to smash the glass ceiling and secure her a permanent academic position, with partial success. Noether blazed the trail of abstract algebra, pioneering today's axiomatic structures such as groups, rings, and fields. She also proved a vital theorem relating the symmetries of laws of physics to conserved quantities, such as energy.

By now the story has moved into the twentieth century. To show that great mathematical ability is not confined to the educated classes of the western world, we follow the life and career of the self-taught Indian genius Srinivasa Ramanujan, who grew up in poverty. His uncanny ability to intuit strange but true formulas was rivalled, if at all, only by giants such as Euler and Carl Jacobi. Ramanujan's concept of proof was hazy, but he could find formulas that no one else would ever have dreamed of. His papers and notebooks are still being mined today for fresh ways of thinking.

Two mathematicians with a philosophical bent return us to the foundations of the subject and its relation to computation. One is Kurt Gödel, whose proof that any axiom system for arithmetic must be incomplete and undecidable demolished Hilbert's programme to prove the opposite. The other is Alan Turing, whose investigations into the abilities of a programmable computer led to a simpler and more natural proof of these results. He is, of course, famous for his codebreaking work at Bletchley Park during World War II. He also proposed the Turing test for artificial intelligence, and after the war he worked on patterns in animal markings. He was gay, and died in tragic and mysterious circumstances.

I decided not to include any living mathematicians, but to end with two recently deceased modern mathematicians: one pure and the other applied (but also unorthodox). The latter is Benoit Mandelbrot, widely known for his work on fractals, geometric shapes that have detailed structure on all scales of magnification. Fractals often model nature far better than traditional smooth surfaces such as spheres and cylinders. Although several other mathematicians worked on structures that we now see as fractal, Mandelbrot made a great leap forward by recognising their potential as models of the natural world. He wasn't a theorem-proof type of mathematician; instead, he had an intuitive visual grasp of geometry, which led him to see relationships and state conjectures. He was also a bit of a showman, and an energetic promoter of his ideas. That didn't endear him to some in the mathematical community, but you can't please everyone.

Finally, I've chosen a (pure) mathematician's mathematician, William Thurston. Thurston, too, had a deep intuitive grasp of geometry, in a broader and deeper sense than Mandelbrot. He could do theorem-proof mathematics with the best of them, though as his career advanced he tended to focus on the theorems and sketch the proofs. In particular he worked in topology, where he noticed an unexpected connection with non-Euclidean geometry. Eventually, this circle of ideas motivated Grigori Perelman to prove an elusive conjecture in topology, due to Poincaré. His methods also proved a more general conjecture of Thurston that provides unexpected insights into all three-dimensional manifolds.

In the final chapter, I'll pick up some of the threads that weave their way through the 25 stories of these astonishing individuals, and explore what they teach us about pioneering mathematicians – who they are, how they work, where they get their crazy ideas, what drives them to be mathematicians in the first place.

For now, however, I'd just like to add two warnings. The first is that I've necessarily been selective. There isn't enough space to provide comprehensive biographies, to survey everything that my trailblazers worked on, or to enter into fine details of how their ideas evolved and how they interacted with their colleagues. Instead, I've tried to offer a representative selection of their most important – or interesting – discoveries and concepts, with enough historical detail to paint a picture of them as people and locate them in their society. For some mathematicians of antiquity, even that has to be very sketchy, because few records about their lives (and often no original documents about their works) have survived.

The second is that the 25 mathematicians I've chosen are by no means the *only* significant figures in the development of mathematics. I made my choices for many reasons – the importance of the mathematics, the intrinsic interest of the area, the appeal of the human story, the historical period, diversity, and that elusive quality, 'balance'. If your favourite mathematician is omitted, the most likely reason is limited space, coupled with a wish to choose representatives that are widely distributed in the three-dimensional manifold whose coordinates are geography, historical period, and gender. I believe that everyone in the book fully deserves inclusion, although one or two may be controversial. I have no doubt at all that many others could have been selected with comparable justification.

1

Do Not Disturb My Circles
Archimedes

Archimedes of Syracuse
Born: Syracuse, Sicily, *c.* 287 BC
Died: Syracuse, *c.* 212 BC

THE YEAR: 1973. The place: Skaramagas naval base, near Athens. All eyes are focused on a plywood mock-up of a Roman ship. Also focused on the ship: the rays of the Sun, reflected from seventy copper-coated mirrors fifty metres away, each a metre wide and half as high again.

Within a few seconds, the ship catches fire.

Ioannis Sakkas, a modern Greek scientist, is recreating a possibly mythical piece of ancient Greek science. In the second century AD the Roman author Lucian wrote that at the Siege of Syracuse, around 214–212 BC, the engineer and mathematician Archimedes invented a device to destroy enemy ships by fire. Whether this device existed, and if so, how it worked, is highly obscure. Lucian's story could just be a reference to the common use of fire arrows or burning rags shot from a catapult, but it's hard to see why this would have been presented as a new invention. In the sixth century, Anthemius of Tralles suggested, in his *Burning Glasses*, that Archimedes had used a huge lens. But in the most prevalent legend, Archimedes used a giant mirror, or possibly an array of mirrors arranged in an arc to form a rough parabolic reflector.

The parabola is a U-shaped curve, well known to Greek geometers. Archimedes certainly knew about its focal property: all lines parallel to the axis, when reflected in the parabola, pass through the same point, called the focus. Whether anyone realised that a parabolic mirror would focus light (and heat) from the Sun in the same way is less certain, because Greek understanding of light was rudimentary. But, as Sakkas's experiment shows, Archimedes wouldn't actually have needed a parabolic arrangement. A lot of soldiers, each armed with a reflecting shield, independently aiming it to direct the Sun's rays towards the same part of the ship, would have been just as effective.

The practicality of what is often called 'Archimedes's heat ray' has been hotly debated. The philosopher René Descartes, a pioneer in optics, didn't believe it could have worked. Sakkas's experiment suggests it might have done, but his fake plywood ship was flimsy, and coated in a tar-based paint, so it would burn easily. On the other hand, in Archimedes's time it was common to coat ships with tar to protect their hulls. In 2005 a bunch of MIT students repeated Sakkas's experiment, eventually setting a wooden mock-up of a ship on fire – but only after focusing the Sun's rays on it for ten minutes while it remained totally stationary. They tried it again for the TV show *Mythbusters* using a fishing boat in San Francisco, and managed to char the wood and produce a few flames, but it didn't ignite. *Mythbusters* concluded that the myth was bust.

✳

Archimedes was a polymath: astronomer, engineer, inventor, mathematician, physicist. He was probably the greatest scientist (to use the modern term) of his age. As well as important mathematical discoveries, he produced inventions that are breathtaking in their scope – the Archimedean screw for raising water, block-and-tackle pulleys to lift heavy weights – and he discovered Archimedes's principle on floating bodies and the law (though not the apparatus, which appeared much earlier) of the lever. He's also credited with a second war-machine, the claw. Allegedly he used this crane-like device at the Battle of Syracuse to lift enemy ships from the water and sink them. The 2005 television documentary *Superweapons of the Ancient World* built its own version of the device, and it worked. Ancient texts contain many other tantalising references to theorems and inventions attributed to Archimedes. Among them is a mechanical planetary calculator, much like the famed Antikythera mechanism of around 100 BC, discovered in a shipwreck in 1900–1901 and only recently understood.

We know very little about Archimedes. He was born in Syracuse (Siracusa), a historic Sicilian city located towards the southern end of the island's east coast. It was founded in either 734 or 733 BC by Greek colonists, supposedly under the semi-mythical Archias when he exiled himself from Corinth. According to Plutarch, Archias had become infatuated with Actaeon, a handsome boy. When his advances were rejected he tried to kidnap the lad, and in the struggle, Actaeon was torn to pieces. His father Melissus's pleas for justice went unanswered, so he climbed to the top of a temple of Poseidon, called upon the god to avenge his son, and flung himself onto the rocks below. A severe drought and famine followed these dramatic events, and the local oracle declared that only vengeance would propitiate Poseidon. Archias got the message, exiled himself voluntarily to avoid being sacrificed, headed for Sicily, and founded Syracuse. Later his past caught up with him anyway when Telephus, who as a boy had also been an object of Archias's desires, killed him.

The land was fertile, the natives friendly, and Syracuse soon became the most prosperous and powerful Greek city in the entire Mediterranean. In *The Sand Reckoner*, Archimedes says that his father was Phidias, an astronomer. According to Plutarch's *Parallel Lives*, he was a distant relative of Hiero II, tyrant of Syracuse. As a young man, Archimedes is thought to have studied in the Egyptian city of Alexandria on the coast of the Nile delta, where he encountered Conon of Samos and Eratosthenes of Cyrene. Among the evidence is his statement that Conon was a friend; also the introductions to his books *The Method of Mechanical Theorems* and the *Cattle Problem* are addressed to Eratosthenes.

There are some tales about his death, too, and we'll come to those in due course.

Archimedes's mathematical reputation rests on those works that have survived – all as later copies. *Quadrature of the Parabola*, which takes the form of a letter to his friend Dositheus, contains 24 theorems about parabolas, the final one giving the area of a parabolic segment in terms of a related triangle. The parabola figures prominently in his work. It's a type of conic section, a family of curves that played a major role in Greek geometry. To create a conic section, use a plane to cut a double cone, formed by joining two identical cones at their tips. There are three main types: the ellipse, a closed oval; the parabola, a U-shaped curve; and the hyperbola, two U-shaped curves back to back.

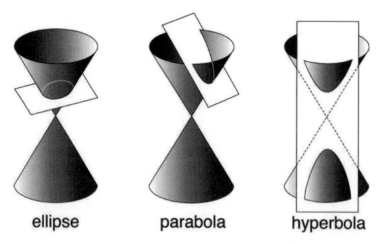

The three main types of conic section.

On Plane Equilibria consists of two separate books. It establishes some fundamental results about what we now call statics: the branch of mechanics that analyses conditions under which a body remains at rest. The further development of this topic underpins the whole of civil engineering, making it possible to calculate the forces that act on the structural elements of buildings and bridges, to ensure that they really do remain at rest, rather than buckling or collapsing.

The first book concentrates on the law of the lever, which Archimedes states as: 'Magnitudes are in equilibrium at distances reciprocally proportional to their weights.' One consequence is that a long lever amplifies a small force. Plutarch tells us that Archimedes dramatised this in a letter to King Heiron: 'Give me a place to stand, and I will move the Earth.' He'd have needed a very long, perfectly rigid, lever, but the main downside of levers is that although the applied force is amplified, the far end of the lever travels a much smaller distance than the force does. Archimedes could have moved the Earth through the same (very tiny) distance just by jumping. Nonetheless, a lever is very effective, and so is a variant that Archimedes also understood, the pulley. When a sceptical Heiron asked him to demonstrate, Archimedes

> fixed accordingly upon a ship of burden out of the king's arsenal, which could not be drawn out of the dock without great labour and many men; and, loading her with many passengers and a full freight, sitting himself the while far off, with no great endeavour, but only holding the head of the pulley in his hand and drawing the cords by degrees, he drew the ship in a straight line, as smoothly and evenly as if she had been in the sea.

The second book is mainly about finding the centre of gravity of various shapes – triangle, parallelogram, trapezium, and segment of a parabola.

On the Sphere and Cylinder contains results of which Archimedes was so proud that he had them inscribed on his tomb. He proved, rigorously, that the surface area of a sphere is four times that of any great circle (such as the equator of a spherical Earth); that its volume is two thirds that of a cylinder fitting tightly round the sphere; and that the area of any segment of the sphere cut off by a plane is the same as the corresponding segment of such a cylinder. His proof used a convoluted method known as exhaustion, which was introduced by Eudoxus to deal with proportions involving irrational numbers, which can't be represented exactly as a fraction. In modern terms, he proved that the surface area of a sphere of radius r is $4\pi r^2$ and its volume is $^4/_3\pi r^3$.

Mathematicians have a habit of presenting their final polished results in beautifully organised fashion, while concealing the often messy and muddled process that led to them. We're fortunate to have some extra insight into how Archimedes made his discoveries about the sphere, recorded in *The Method of Mechanical Theorems*. This work was long thought to be lost, but in 1906 the Danish historian Johan Heiberg discovered an incomplete copy, the Archimedes palimpsest. A palimpsest is a text that was rubbed out or washed off in antiquity to allow the parchment or paper to be reused. The works of Archimedes were collected together by Isidorus of Miletus around 530 in Constantinople (modern Istanbul), capital of the Byzantine Empire. They were copied in 950 by some Byzantine scribe, at a time when Leo the Geometer was running a mathematical school studying Archimedes's works. The manuscript made its way to Jerusalem, where in 1229 it was disassembled, washed (not very effectively), folded in half, and rebound to make a 177-page Christian liturgy.

In the 1840s the biblical scholar Constantin von Tischendorf came across this text, by now back in a Greek orthodox library in Constantinople, and noticed faint traces of Greek mathematics. He took one page away and deposited it in Cambridge University Library. In 1899 Athanasios Papadopoulos-Kerameus, cataloguing the Library's manuscripts, translated part of it. Heiberg realised it was by Archimedes, and tracked the page back to Constantinople, where he was allowed to photograph the entire document. He then transcribed it, publishing the results between 1910 and 1915, and Thomas Heath translated the text into English. After a complicated series of events, including an auction contested by a lawsuit over ownership, it was sold to an anonymous American for two million dollars. The new owner made it available for study, and it was subjected to a variety of digital imaging techniques to bring out the underlying text.

The technique of exhaustion requires advance knowledge of the answer, and scholars had long wondered how Archimedes guessed the rules for the area and volume of a sphere. *The Method* provides an explanation:

> Certain things first became clear to me by a mechanical method, although they had to be proved by geometry afterwards because their investigation by the said method did not furnish an actual proof. But it is of course easier, when we have previously acquired, by the method, some knowledge of the questions, to supply the proof than it is to find it without any previous knowledge.

Archimedes imagines hanging a sphere, a cylinder, and a cone on a balance, and then cutting them into infinitely thin slices, which are redistributed in a way that keeps the balance level. He then uses the law of the lever to relate the three volumes (those of the cylinder and cone were known) and deduces the required quantities. It has been suggested that Archimedes was pioneering the use of actual infinities in mathematics. This may be reading too much into an obscure document, but it's clear that *The Method* anticipates some ideas of calculus.

Archimedes's other works illustrate how diverse his interests were. *On Spirals* proves some fundamental results about lengths and areas related to the Archimedean spiral, the curve described by a point moving at uniform speed along a line rotating at uniform speed. *On Conoids and Spheroids* studies the volumes of segments of solids formed by rotating conic sections about an axis.

On Floating Bodies is the earliest work in hydrostatics, equilibrium positions of floating objects. It includes Archimedes's principle: a body immersed in a liquid is subjected to a buoyancy force equal to the weight of fluid displaced. This principle is the subject of a famous anecdote in which Archimedes is asked to devise a method to determine whether a votive crown made for King Hiero II is truly made of gold. Sitting in his bath, he is suddenly struck with inspiration, becoming so excited that he rushes off down the street crying '*Eureka*!' (I've found it!) – omitting to get dressed first.

Public nudity would not have been particularly scandalous in ancient Greece, mind you. The technical high point of the book is a condition for a floating paraboloid to be stable, a forerunner of basic ideas in naval architecture on the stability and capsizing of ships.

Measurement of a Circle applies the method of exhaustion to prove that the area of a circle is half the radius times the circumference – πr^2 in modern terms. To prove this, Archimedes inscribes and circumscribes regular polygons with 6, 12, 24, 48, and 96 sides. By considering the 96-gons, he proves a result equivalent to an estimate for the value of π: it lies between $3^1/_7$ and $3^{10}/_{71}$.

The *Sand Reckoner* is addressed to Gelo II, tyrant of Syracuse, the son of Heiro II. This adds evidence that Archimedes had royal connections. He explains its objective:

> There are some, king Gelo, who think that the number of the sand is infinite in multitude ... But I will try to show you ... that, of the numbers named by me and given in the work which I sent to Zeuxippus, some exceed not only the number of the mass of sand equal in magnitude to the Earth filled up, but also that of the mass equal in magnitude to the universe.

Here Archimedes is promoting his new system for naming large numbers by combating the common misuse of the term 'infinite' to mean 'very large'. He has a clear sense of the distinction. His text combines two main ideas. The first is an extension of the standard Greek number words to allow much larger numbers than a myriad myriad (100 million, 10^8). The second is an estimate for the size of the universe, which he bases on the heliocentric (Sun-centred) theory of Aristarchus. His final result is that, in today's notation, it would take at most 10^{63} sand grains to fill the universe.

There's a long recreational tradition in mathematics, featuring games and puzzles. Sometimes these are just fun, and sometimes they're light-hearted problems that illuminate more serious concepts. The *Cattle Problem* raises questions that are still studied today. In 1773 Gotthold Lessing, a German librarian, came across a Greek manuscript: a 44-line poem inviting the reader to calculate how many cattle there are in the Sun god's herd. The title of the poem presents it as a letter from Archimedes to Eratosthenes. It begins:

> Compute, O friend, the number of the cattle of the Sun which once grazed upon the plains of Sicily, divided according to colour into four herds, one milk-white, one black, one dappled and one yellow. The number of bulls is greater than the number of cows, and the relations between them are as follows.

It then lists seven equations along the lines of

$$\text{white bulls} = \left(\tfrac{1}{2} + \tfrac{1}{3}\right) \text{ black bulls} + \text{yellow bulls}$$

and continues:

> If thou canst give, O friend, the number of each kind of bulls and cows, thou art no novice in numbers, yet cannot be regarded as of high skill. Consider, however, the following additional relations between the bulls of the Sun:
> White bulls + black bulls = a square number,
> Dappled bulls + yellow bulls = a triangular number.
> If thou hast computed these also, O friend, and found the total number of cattle, then exult as a conqueror, for thou hast proved thyself most skilled in numbers.

Square numbers are 1, 4, 9, 16, and so on, found by multiplying a whole number by itself. Triangular numbers are 1, 3, 6, 10 and so on, formed by adding consecutive whole numbers – for instance, $10 = 1 + 2 + 3 + 4$. These conditions constitute what we now call a system of Diophantine equations, named after Diophantus of Alexandria, who wrote about them around AD 250 in *Arithmetica*. The solution must be given in whole numbers, since the Sun god would be unlikely to have half a cow in his herd.

The first set of conditions leads to an infinite number of possible solutions, the smallest giving 7,460,514 black bulls and comparable numbers of the other animals. The supplementary conditions select among those solutions, and lead to a type of Diophantine equation known as the Pell equation (Chapter 6). This asks for integers x and y such that $nx^2 + 1 = y^2$ where n is a given integer. For example, when $n = 2$ the equation is $2x^2 + 1 = y^2$, with solutions such as $x = 2$, $y = 3$ and $x = 12$, $y = 17$. In 1965 Hugh Williams, R.A. German, and Charles Zarnke found the smallest solution consistent with the two extra conditions, using two IBM computers. It's approximately $7 \cdot 76 \times 10^{206544}$.

There's no way Archimedes could have found this number by hand, and there's no evidence that he had anything to do with the problem, beyond the poem's title. The cattle problem still attracts the attention of number theorists, and has inspired new results on the Pell equation.

The historical record of Archimedes's life is flimsy, but we know a little more about his death, assuming any of the tales is accurate. They probably contain at least a grain of truth.

In the Second Punic War, around 212 BC, the Roman general Marcus Claudius Marcellus besieged Syracuse, capturing it after two years. Plutarch relates that the elderly Archimedes was looking at a geometric diagram in the sand. The general sent a soldier to tell Archimedes to meet him, but the mathematician protested that he hadn't finished working on his problem. The soldier lost his temper and killed Archimedes with his sword; the sage's last words were allegedly 'Do not disturb my circles!' Knowing mathematicians, I find this entirely plausible, but Plutarch gives another version in which Archimedes tries to surrender to a soldier, who thinks the mathematical instruments he is carrying are valuable and slaughters him to steal them. In both versions, Marcellus was somewhat peeved at the death of this revered mechanical genius.

Archimedes's tomb was decorated with a sculpture depicting his favourite theorem, from *On the Sphere and Cylinder*: a sphere inscribed in a cylinder has two thirds its volume and the same surface area. More than a century after Archimedes's death, the Roman orator Cicero was a quaestor (state-appointed auditor) in Sicily. Hearing of the tomb, he eventually found it in a dilapidated state near the Agrigentine gate in Syracuse. He ordered its restoration, which let him read some of its inscriptions, including a diagram of the sphere and cylinder.

Today, the location of the tomb is unknown, and nothing appears to have survived. But Archimedes lives on through his mathematics, much of it still important more than two thousand years later.

2

Master of the Way
Liu Hui

Liu Hui
Flourished: Cao Wei, China, third century AD

THE *ZHOU BI SUAN JING* (Arithmetical Classic of the Gnomon and the Circular Paths of Heaven) is the most ancient known Chinese mathematical text, dating from the period of the Warring States, 400–200 BC. It opens with a neat piece of educational propaganda:

> Long ago, Rong Fang asked Chen Zi 'Master, I have recently heard something about your Way. Is it really true that your way is able to comprehend the height and size of the Sun, the area illuminated by its radiance, the amount of its daily motion, the figures for its

greatest and least distances, the extent of human vision, the limits of the four poles, the constellations into which the stars are ordered, and the length and breadth of heaven and Earth?'

'It is true,' said Chen Zi.

Rong Fang asked, 'Although I am not intelligent, Master, I would like you to favour me with an explanation. Can someone like me be taught this Way?'

Chen Zi replied, 'Yes. All things can be attained to you by mathematics. Your ability in mathematics is sufficient to understand such matters if you sincerely give repeated thought to them.'

The book goes on to derive a figure for the distance from the Earth to the Sun, using geometry. Its cosmological model was primitive: a flat Earth beneath a plane circular sky. But its mathematics was quite sophisticated. Essentially it used the geometry of similar triangles, applied to shadows cast by the Sun.

The *Zhou Bi* shows the advanced state of Chinese mathematics around the time of the Greek Hellenistic period, from the death of Alexander the Great in 323 BC to 146 BC when the Republic of Rome added Greece to its empire. This period was the peak of ancient Greek intellectual dominance; the time of most of the great geometers, philosophers, logicians, and astronomers of the classical world. Even under Roman dominion, Greece continued to make cultural and scientific advances until about AD 600, but the centres of mathematical innovation moved to China, Arabia, and India. The cutting edge of mathematical progress didn't return to Europe until the Renaissance, although the 'dark ages' weren't as dark as they're sometimes painted, and lesser advances were made in Europe too.

The Chinese advances were stunning. Until recently most histories of mathematics adopted a Eurocentric viewpoint and ignored them, until George Gheverghese Joseph wrote about the early mathematics of the Far East in *The Crest of the Peacock*. Among the greatest of the ancient Chinese mathematicians was Liu Hui. A descendant of the Marquis of Zixiang of the Han dynasty, he lived in the state of Cao Wei during the Three Kingdoms period. In 263, he edited and published a book with solutions to mathematical problems presented in the famous Chinese mathematics book *Jiuzhang Suanshu* (Nine Chapters on the Mathematical Art).

His works include a proof of Pythagoras's Theorem, theorems in solid geometry, an improvement on Archimedes's approximation to π, and a systematic method for solving linear equations in several unknowns. He also wrote about surveying, with especial application to astronomy. He probably visited Luoyang, one of the four ancient capitals of China, and measured the Sun's shadow.

Evidence for the earliest history of China comes from a few later texts, such as the Han dynasty scribe Sima Qian's vast *Records of the Grand Historian* (around 110 BC) and the *Bamboo Annals*, a historical chronicle written on slips of bamboo, buried in the grave of King Xiang of Wei in 296 BC and dug up again in AD 281. According to these sources, Chinese civilisation began in the third millennium BC with the Xia Kingdom. Written records start with the Shang dynasty, which ruled from 1600–1046 BC and left the earliest evidence of Chinese counting in the form of oracle bones – marked bones used for fortune-telling. A successful invasion by the Zhou led to a more stable state with a feudal structure, which began to fall apart three centuries later as other groups tried to muscle in.

By 476 BC virtual anarchy ruled, a period known as the Warring States that lasted over two centuries. The *Zhou Bi* was written during these turbulent times. Its main mathematical contents are what we now call Pythagoras's Theorem, fractions, and arithmetic; it also includes a lot of astronomy. Pythagoras's Theorem is presented in a conversation between Duke Chou Kung and the noble Shang Kao. Their discussion of right triangles leads to a statement of the famous theorem and a geometric proof. For a time historians thought that this discovery beat Pythagoras by half a millennium. The general view today is that it was an independent discovery, predating Pythagoras, but not by much.

An important successor from the same general period is the aforementioned *Jiuzhang*, which contains a wealth of material such as the extraction of roots, solution of simultaneous equations, areas and volumes, and again right triangles. A commentary by Chang Heng in AD 130 gives the approximation $\pi\sqrt{10}$. Chao Chun Chin's commentary on *Zhou Bi* some time in the third century AD added a method for solving quadratic equations. The most influential development from *Jiuzhang* was made by China's greatest mathematician of antiquity, Liu Hui in AD 263. He introduced the book with an explanation:

> In the past, the tyrant Qin burnt written documents, which led to the destruction of classical knowledge. Later, Zhang Cang, Marquis of Peiping, and Geng Shouchang, Vice-President of the Ministry of Agriculture, both became famous through their talent for calculation. Because the ancient texts had deteriorated, Zhang Cang and his team produced a new version removing the poor parts and filling in the missing parts. Thus, they revised some parts, with the result that these were different from the old parts.

In particular, Liu Hui provided proofs that the book's methods work, using techniques that today we wouldn't consider rigorous, akin to those of Archimedes in *The Method*. And he supplied additional material on surveying, also published separately as *Haidao Suanjing* (Sea Island Mathematical Manual).

The first chapter of the *Jiuzhang* explains how to calculate the areas of fields of various shapes, such as rectangles, triangles, trapeziums, and circles. Its rules are correct, except for the circle. Even here the *recipe* is right: multiply the radius by half the circumference. However, the circumference is calculated as 3 times the diameter, in effect taking $\pi = 3$. As a practical matter, the rule underestimates the area by less than 5 per cent.

Late in the first century BC the ruler Wang Mang instructed the astronomer and calendar-maker Liu Hsing to come up with a standard measure for volume. Liu Hsing made a very accurate cylindrical bronze vessel, to act as a standard reference measure. Thousands of copies were used all over China. The original vessel is now in a museum in Beijing, and its dimensions have led some to suggest that Liu Hsing in effect used a value for π somewhere around 3·1547. (Quite how the figure can be obtained to this degree of accuracy by measuring a bronze pot is unclear, to me at least.) The *Sui Shu* (official history of the Sui dynasty) contains a statement equivalent to Liu Hsing having found a new value for π. Liu Hui remarked that around the same time, the court astrologer Chang Heng proposed taking π to be the square root of 10, which is 3·1622. Clearly improved values for π were in the air.

In his commentary on the *Jiuzhang*, Liu Hui points out that the traditional 'π = 3' rule is wrong: instead of the circumference of the circle, it gives the perimeter of an inscribed hexagon, which is visibly smaller. He then calculated a more accurate value for the circumference (and, implicitly, for π). In fact, he went further, describing a computational method to estimate π to arbitrary accuracy. His approach was similar to that of Archimedes: approximate the circle by regular polygons with 6, 12, 24, 48, 96, ... sides. In order to apply the method of exhaustion, Archimedes used one sequence of approximating polygons inscribed in the circle, and a second sequence fitting outside it. Liu Hui used only inscribed polygons, but at the end of the calculation he gave a geometric argument to place both lower and upper bounds on the true value of π. This method gives arbitrarily accurate approximations to π using nothing more difficult than square roots. These can be calculated systematically; the method is laborious but no more complex than long multiplication. A skilled arithmetician could obtain ten decimal places of π in a day.

Later, around AD 469, Tsu Ch'ung Chih extended the calculation to show that

$$3 \cdot 1415926 < \pi < 3 \cdot 1415927$$

The result was recorded, but his method, which may have been explained in his lost work *Su Shu* (Method of Interpolation), was not. It could have been done by continuing Liu Hui's calculation, but the book's title suggests it involved estimating a more accurate value from a pair of approximations, one too small and the other too big. Methods like that can be found in mathematics right up to the present day. Not so long ago they were taught in schools, for use with tables of logarithms. Tsu came up with two simple fractions approximating π : the Archimedean 22/7, accurate to two decimal places, and 355/113, accurate to six decimal places. The first is widely used today, and the second is well known to mathematicians.

❋

One reconstruction of Liu Hui's proof of Pythagoras's Theorem, based on the instructions in his book, is an ingenious and unusual dissection. The right triangle is shown in black. The square on one side is split in two by a diagonal (light grey). The other square is cut into five pieces: one small square (dark grey), a pair of symmetrically arranged triangles the same shape and size as the original right triangle (medium grey) and a pair of symmetrically arranged triangles filling the remaining space (white). Then all seven pieces are assembled to make the square on the hypotenuse.

Other, simpler dissections can also be used to prove this theorem.

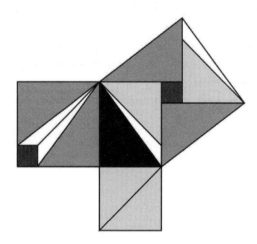

Possible reconstruction of Liu Hui's proof of Pythagoras's Theorem.

The ancient Chinese mathematicians were every bit as capable as their Greek contemporaries, and the course of Chinese mathematics after Liu Hui's period includes many discoveries that predate their appearance in European mathematics. For example, the estimates for π found by Liu Hui and Tsu Ch'ung Chih were not bettered for a thousand years.

Joseph examines whether some of their ideas might have been transmitted to India and Arabia alongside trade goods, and thence possibly even to Europe. If so, the later European rediscoveries might not have been entirely independent. There were Chinese diplomats in

India in the sixth century, and Chinese translations of Indian mathematics and astronomy books were made in the seventh century. As regards Arabia, the Prophet Muhammad issued a *hadith* – a pronouncement with religious significance – saying 'Seek learning, though it be as far away as China.' In the fourteenth century Arab travellers report formal trade links with China, and the Moroccan traveller and scholar Muhammad ibn Battuta wrote about Chinese science and technology, as well as culture, in *Rihla* (Journey).

We know that ideas from India and Arabia made their way to medieval Europe, as the next two chapters illustrate. So it's by no means impossible that Chinese knowledge did likewise. The Jesuit presence in China in the seventeenth and eighteenth centuries inspired some of Leibniz's philosophy, by way of Confucius. There may well have been a complex network, transmitting mathematics, science, and much more, between Greece, the Middle East, India, and China. If so, the conventional history of western mathematics may require some revision.

3

Dixit Algorismi
Muhammad al-Khwarizmi

Muhammad ibn Musa al-Khwarizmi
Born: Khwarizm (modern-day Khiva), Persia, *c.* 780
Died: *c.* 850

AFTER THE DEATH of the Prophet Muhammad in 632, control of the Islamic world passed to a series of caliphs. In principle, caliphs were chosen on merit, so the caliphate's power system wasn't exactly a monarchy. However, the caliph was very much in charge. By 654 under Uthman, the third caliph, the caliphate had become the largest empire the world had ever seen. Its territory (in present-day geography) included the Arabian Peninsula, North Africa from Egypt through Libya to eastern Tunisia, the Levant, the Caucasus, and much of Central Asia from Iran into Pakistan, Afghanistan, and Turkmenistan.

The first four caliphs constituted the Rashidun caliphate, succeeded by the Umayyad dynasty, which was in turn succeeded by the Abbasid dynasty, which overthrew the Umayyads with Persian assistance. The centre of government, originally in Damascus, moved to Baghdad, a city founded by Caliph al-Mansur in 762. Its location, close to Persia, was in part dictated by the need to rely on the services of Persian administrators, who understood how the various regions of the Islamic Empire interacted. The position of vizier was created, allowing the caliph to delegate administrative responsibility; the vizier in turn delegated local matters to regional emirs. The caliph's position slowly became that of figurehead, with the real power residing in the vizier, but the early Abbasid caliphs exerted considerable control.

Around 800 Harun al-Rashid founded the *Bayt al-Hikma*, or House of Wisdom, a library in which writings from other cultures were translated into Arabic. His son al-Ma'mun pushed the project through to completion, assembling a huge collection of Greek manuscripts and a large number of scholars. Baghdad became a centre for science and trade, attracting scholars and merchants from places as distant as China and India. Among them was Muhammad ibn Musa al-Khwarizmi, a key figure in the history of mathematics.

Al-Khwarizmi was born in or near Khwarizm in Central Asia, now Khiva in Uzbekistan. He did his main work under al-Ma'mun, helping to keep alive the knowledge that Europe was fast losing. He translated key Greek and Sanskrit manuscripts, made his own advances in science, mathematics, astronomy, and geography, and wrote a series of books that we would now describe as scientific bestsellers. *On Calculation with Hindu Numerals*, written around 825, was translated into Latin as *Algoritmi de Numero Indorum*, and it almost singlehandedly spread the news of this amazing new way to do arithmetic to medieval Europe. Along the way, Algoritmi became Algorismi, and methods for calculating with these numerals were called algorisms. In the eighteenth century, the word changed to algorithm.

His *al-Kitab al-mukhtasar fi hisab al-jabr wa-l-muqabala* (The Compendious Book on Calculation by Completion and Balancing), written around 830, was translated into Latin in the twelfth century by Robert of Chester as *Liber Algebrae et Almucabola*. As a result, *al-jabr*, Latinised to 'algebra', became a word in its own right. It now refers to the use of symbols such as x and y for unknown quantities, together with methods for finding those unknowns by solving equations, but the book doesn't use symbols.

The *Algebra* came about when Caliph al-Ma'mun encouraged al-Khwarizmi to write a popular book about calculation. Its author describes its purpose as explaining

> what is easiest and most useful in arithmetic, such as men
> constantly require in cases of inheritance, legacies, partition,
> lawsuits, and trade, and in all their dealings with one another, or
> where the measuring of lands, the digging of canals, geometrical
> computations, and other objects of various sorts and kinds are
> concerned.

That doesn't sound much like an algebra book. In fact, algebra occupies only a small part. Al-Khwarizmi begins by explaining numbers in very simple terms – units, tens, hundreds – on the grounds that 'when I consider what people generally want in calculating, I find that it always is a number'. It wasn't a learned treatise aimed at scholars; it was a popular mathematics book, one of the educational kind that tries to teach the general reader as well as inform them. That's what the caliph wanted, and that's what he got. Al-Khwarizmi didn't consider his book to be at the frontiers of research mathematics. But that's how we now see the part on *al-jabr*. This is the deepest section of the book: a systematic development of methods for solving equations in some unknown quantity.

Al-jabr, usually translated as 'completion', refers to the addition of the same term to each side of the equation, with the aim of simplifying it. *Al-muqabala*, 'balancing', refers to the removal of a term ·on one side of the equation to the other side (but with the opposite sign) and to the cancellation of like terms on both sides.

For example, if the equation, expressed in modern symbolic notation, is

$$x - 3 = 7$$

then *al-jabr* allows us to add 3 to each side, obtaining

$$x = 10$$

which in this case solves the equation. If it's

$$2x^2 + x + 6 = x^2 + 18$$

then *al-muqabala* lets us move 6 on the left side to the right, as long as we subtract it, which yields

$$2x^2 + x = x^2 + 12$$

A second *al-muqabala* lets us move x^2 on the right over to the left and subtract *that*, getting

$$x^2 + x = 12$$

which is simpler, though not yet the answer.

I repeat that al-Khwarizmi *doesn't use symbols*. The father of algebra didn't actually do what most of us think of as algebra. He stated everything verbally. Specific numbers were *units*, the unknown quantity that we call x was *root*, and our x^2 was *square*. The previous equation would read:

square plus *root* equals twelve *units*

without any symbols. So the next job is to explain how to go from this sort of equation to the answer. Al-Khwarizmi classifies equations into six types, a typical case being 'squares and roots equal to numbers', such as $x^2 + x = 12$.

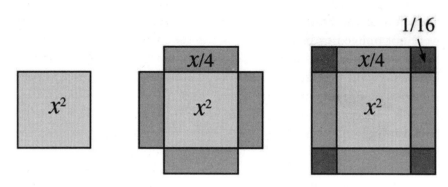

Geometric solution of 'squares and roots equal to numbers'.

He then proceeds to analyse each type in turn, solving the equation using a mixture of algebraic and geometric methods. Thus to solve the equation $x^2 + x = 12$, al-Khwarizmi draws a square to represent x^2 (left-hand picture). To add the root x, he adjoins four rectangles, each of sides x and $^1/_4$ (middle picture). The resulting shape leads to the idea of 'completing the square' by adjoining four small squares, each of side $^1/_4$ and area $^1/_{16}$. So he adds $4 \times {}^1/_{16} = {}^1/_4$ to the left-hand side of the equation as well (right-hand picture). By the rule of *al-jabr*, he must also add $^1/_4$ to the right-hand side of the equation, which becomes $12^1/_4$. Now

$$\left(x + {}^1/_2\right)^2 = 12^1/_4 = 49/4 = (7/2)^2$$

Take the square root to get

$$x + {}^1/_2 = 7/2$$

so $x = 3$. Today we would also take the negative square root, $-7/2$, and obtain a second solution $x = -4$. Negative numbers were starting to be understood in al-Khwarizmi's day, but he doesn't mention them.

Both the Babylonians and the Greeks would have understood this approach, because they'd already done much the same. In fact, there's some controversy about whether al-Khwarizmi was aware of Euclid's *Elements*. He should have been, because al-Hajjaj, another scholar at the House of Wisdom, had translated Euclid into Arabic when al-Khwarizmi was a young man. But on the other hand, the main job of the House of Wisdom was *translation*, and its workers were not obliged to read the works translated by their fellows. Some historians argue that al-Khwarizmi's geometry is not presented in Euclid's style, suggesting a lack of familiarity. But, I repeat, the *Algebra* is a popular mathematics book, so it wouldn't follow Euclid's axiomatic style, even if al-Khwarizmi knew Euclid backwards. In any case, completing the square goes right back to the Babylonians, and was available from many sources.

Why, then, do many historians consider al-Khwarizmi the father of algebra? Especially when he uses no symbols? There's a strong competitor, the Greek Diophantus. His *Arithmetica*, a series of books on the solution of equations in whole or rational numbers written around 250, did use symbols. One answer is that Diophantus's main interest was number theory, and his symbols are little more than abbreviations. A deeper point, which I find more convincing, is that al-Khwarizmi often, though not always, provides general *recipes*. The typical presentational style of his predecessors was to use an example with specific numbers, tell you how to solve that, and leave you to infer the general rule. So the upshot of the geometric argument above might have been presented as 'take 1, divide it by 2 to get $1/2$, square that to get $1/4$, then add $1/4$ to both sides', leaving the reader to infer that the general rule is to replace the initial 1 by half the coefficient of x, square that, and add the result to both sides, and so on. This level of generality would of course have been made clear by the tutor, and reinforced by making the student work out lots of other examples.

Sometimes al-Khwarizmi appears to do the same, but he tends to be more explicit about the rule that's being applied. So the deeper reason for crediting him with the invention of algebra is that he focuses more on the generalities of manipulating algebraic expressions than on the numbers they represent. For example, he states a rule for expanding a product

$$(a + bx)(c + dx)$$

in terms of the square x^2, the root x, and numbers. We would write his rule symbolically as

$$ac + (ad + bc)x + (bd)x^2$$

and this is what he states, verbally, without using specific numbers for *a*, *b*, *c*, or *d*. He is telling his readers how to manipulate *general* expressions in numbers, roots, and squares. They're not thought of as coded versions of an unknown number, but as a new kind of mathematical object, which you can calculate with even when you don't know the actual numbers. It's this step towards abstraction – if we accept it as such – that underpins the claim that al-Khwarizmi invented algebra. There's nothing like it in *Arithmetica*.

Other topics in the book are more prosaic: rules for areas and volumes of figures such as rectangles, circles, cylinders, cones, and spheres. Here al-Khwarizmi follows the treatment in Hindu and Hebrew texts, and nothing in it looks much like Archimedes or Euclid. The book ends with more down-to-earth matters: an extensive treatment of Islamic rules for the inheritance of property, requiring division in various proportions, but nothing more complicated mathematically than linear equations and basic arithmetic.

Al-Khwarizmi's most influential work, at the time he wrote it and for several hundred years afterwards, was *On Calculation with Hindu Numerals*, which, as already remarked, gave us the word 'algorithm'. The phrase 'dixit Algorismi' – 'thus spoke al-Khwarizmi' – was a potent argument in any mathematical dispute. The master has spoken: heed his words.

Hindu numerals, of course, are the early versions of decimal notation, in which any number can be written as a sequence of ten symbols, 0 1 2 3 4 5 6 7 8 9. As the book's title indicates, al-Khwarizmi gave credit to the Hindu mathematicians, but so great was his influence in medieval Europe that the idea has become known as Arabic numerals (or sometimes Hindu-Arabic, which is still unfair to the Hindus.) The Arabic world's main contribution was to invent its own number symbols, related to but distinct from the Indian ones, and to disseminate the notation and encourage its use. The symbols for the ten digits have changed repeatedly with the passage of time, and different regions of the modern world still use different symbols.

Today an algorithm is a step-by-step procedure that computes some specific quantity, or produces some specific output, with a guarantee that it gets the right answer and stops. 'Keep trying numbers at random until one works' isn't an algorithm: if it gets an answer, it's correct, but it might keep trying forever without finding anything. For an early example of an algorithm, recall that a prime number has no factors other than itself and 1. The first few primes are 2, 3, 5, 7, 11, 13. Any other positive whole number greater than 1 is said to be composite. For example, 6 is composite because $6 = 2 \times 3$. The number 1 is deemed to be special, and is called a unit in this context. The sieve of Eratosthenes, from around 250 BC, is an algorithm for writing down all prime numbers up to a given limit, as follows. Start by listing positive whole numbers up to that limit. Remove all multiples of 2 except 2, then remove all multiples of the next surviving number 3, aside from 3 itself, then do the same for the next surviving number 5, and so on. After a number of steps that's less than the chosen limit, the process ends by listing precisely the prime numbers up to that limit.

Algorithms have become central to modern life, because computers are machines that run algorithms. Algorithms post cute cat videos to the internet, calculate your credit rating, decide which books to try to sell you, implement billions of currency and stock market trades every second, and try to steal your online banking password. Ironically, the place where algorithms are the most significant feature of al-Khwarizmi's work is not in *On Calculation with Hindu Numerals*, though every method for arithmetical calculation is of course an algorithm. It's his algebra book, whose claim to fame is the specification of general procedures to solve equations. Those procedures are algorithms, and that's what makes them important.

Al-Khwarizmi wrote on geography and astronomy as well as mathematics. His *Kitab surat al-ard* (Book of the Description of the Earth) of 833 updates the previous standard work on this topic,

Ptolemy's *Geography* of around 150. This is a kind of do-it-yourself atlas of the then known world: outlines of the continents on three alternative types of coordinate grid, with instructions for where to put major cities and other prominent features. It also discusses basic principles of map-making. His revision expanded the list to 2402 locations and corrected some of Ptolemy's data, in particular reducing his overestimate for the length of the Mediterranean. While Ptolemy showed the Atlantic and Indian oceans as seas surrounded by land, al-Khwarizmi left them unbounded.

Zij al-Sindhind (Astronomical Tables of the Sindhind), which dates from around 820, contains over a hundred astronomical tables, mainly taken from the works of Indian astronomers. They include tables of the motion of the Sun, Moon, and the five planets, together with tables of trigonometric functions. It is thought that he also wrote on spherical trigonometry, which is important in navigation. *Risala fi istikhraj ta'rikh al-yahud* (Extraction of the Jewish Era) is about the Jewish calendar, and discusses the Metonic cycle, a 19-year period that is very close to a common multiple of the solar year and the lunar month. In consequence, solar and lunar calendars, which tend to diverge as time passes, almost come back into alignment every 19 years. It's named after Meton of Athens, who introduced it in 432 BC.

Alongside the mathematicians of ancient China (Chapter 2) and India (Chapter 4), al-Khwarizmi's achievements add to the weight of evidence that during the Middle Ages, when Europe's science mostly stagnated, the centre of scientific and mathematical advances moved to the Far and Middle East. Eventually, during the Renaissance, Europe woke up again, as we'll see in Chapter 5. Al-Khwarizmi had blazed a new trail, and mathematics would never look back.

4

Innovator of the Infinite
Madhava of Sangamagrama

Irinnarappilly (or Irinninavalli) Madhava
Born: Sangamagrama, Kerala, India 1350
Died: India, 1425

'THE WATER IN HURRICANE RITA weighed as much as 100 million elephants.' Today the media often use elephants as a unit of weight, not to mention Belgium and Wales as measures of area, Olympic swimming pools as measures of volume, and London buses for length or height. So what do you make of this?

> Gods (33), eyes (2), elephants (8), snakes (8), fires (3) qualities (3), vedas (4), naksatras (27), elephants (8), and arms (2) – the wise say that this is the measure of the circumference when the diameter of a circle is 900,000,000,000.

Anything spring to mind? Actually, it's a translation of a poem about π, written around 1400 by Madhava of Sangamagrama, probably the greatest of the medieval Indian mathematician-astronomers. The gods, elephants, snakes, and so on, are number symbols, which would have been drawn as small pictures. Collectively (work backwards through the list) they represent the number

$$282, 743, 388, 233$$

which, divided by 900 billion, gives

$$3 \cdot 141592653592222\ldots$$

This should look more familiar. The ratio concerned is the geometric definition of π, which is

$$3 \cdot 141592653589793\ldots$$

The two figures agree to 11 decimal places (rounding up the 589 to get 59 for the tenth and eleventh decimal places). At the time this was one of the best approximations known. By 1430, the Persian mathematician Jamshid al-Kashi had broken the record with 16 decimal places in *Miftah al-hisab* (The Key to Arithmetic).

Some of Madhava's astronomical texts have survived, but his mathematical work is known only through later commentaries. The perennial problem of giving the great founder and master credit for results found by his intellectual descendants (so that, for example, anything discovered by a member of the Pythagorean cult is attributed by default to Pythagoras) means that we can't be sure exactly which results were discovered by Madhava. In what follows, I'll take his successors at their word.

His greatest achievement was to introduce infinite series, thereby taking early steps towards analysis. He found what in the west is known as Gregory's series for the inverse tangent function, leading to expressions for π as infinite series. His most impressive discoveries are infinite series for the trigonometric sine and cosine functions, which were found in the west over two hundred years later, by Newton.

Little is known of Madhava's life. He lived in the village of Sangamagrama, and this is conventionally appended to his name to distinguish him from other Madhavas, such as the astrologer Vidya Madhava. The village had a temple devoted to a god of the same name. It is believed to have been located near the modern Brahmin village of Irinjalakuda. This is close to Cochin in the state of Kerala, a long thin region near the southern tip of India, sandwiched between the Arabian Sea along its western coast and the Western Ghat mountains to the east. In late medieval times Kerala was a hotbed of mathematical research. Most earlier Indian mathematics originated further north, but for some unknown reason Kerala underwent an intellectual revival. Mathematics was generally seen as a branch of astronomy in ancient India, and Madhava founded the Kerala school of astronomy and mathematics.

This included a number of unusually proficient mathematicians. Parameshvara was a Hindu astronomer who used observations of eclipses to check the accuracy of the computational methods of the day. He wrote at least 25 manuscripts. Kelallur Nilakantha Somayaji wrote a major astronomy text *Tantrasamgraha* in 1501, consisting of 432 Sanskrit verses organised as eight chapters. In particular, it includes his modifications of the great Indian mathematician Aryabhata's theory of the motion of Mercury and Venus. He also wrote an extensive commentary *Aryabhatiya Bhasya* on other work of Aryabhata, in which he discussed algebra, trigonometry, and infinite series for trigonometric functions. Jyesthadeva wrote *Yuktibhasa*, a commentary on *Tantrasamgraha* that added proofs of its main results. Some consider it the first calculus text. Melpathur Narayana Bhattathir, a mathematical linguist, extended Panini's axiomatic system of 3959 rules for Sanskrit grammar in *Prkriya-sarvawom*. He is celebrated for *Narayaneeyam*, a song of praise to Krishna still in use today.

Trigonometry, the use of triangles for measurement, goes back to the ancient Greeks, especially Hipparchus, Menelaus, and Ptolemy.

There are two main types of application: surveying and astronomy. (Later, navigation was added to the list.) The essential point is that distances are often hard to measure directly (in the case of astronomical bodies, impossible), but angles can be measured whenever there's a clear line of sight. Trigonometry makes it possible to deduce the lengths of the sides of a triangle from its angles, provided at least one length is known. In surveying, one carefully measured accessible baseline and a lot of angles lead to an accurate map, and the same goes for astronomy, with tactical differences.

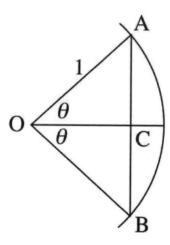

Let AB be an arc of a circle of radius 1, centre O. The chord of angle AOB (whose size is 2θ) is the length of AB. The sine of angle AOC (whose size is θ) is the length of AC. The cosine of θ is the length of OC, and the tangent is AC/OC.

The Greeks worked with the chord of an angle, see the illustration above. Hipparchus produced the first table of chords in 140 BC, and used it for both plane and spherical trigonometry. The latter is about triangles formed by arcs of great circles on a sphere, and it's essential for astronomy because stars and planets appear to lie on the celestial sphere, an imaginary sphere centred on the Earth. More precisely, the directions to these bodies correspond to points on any such sphere. In the second century Ptolemy included tables of chords in his *Almagest*, and his results were widely used for the next 1200 years.

The mathematicians of ancient India built on the Greek work to make major advances in trigonometry. They found it more convenient to use not chords, but the closely related sine (sin) and cosine (cos) functions, which we still do today. Sines first appeared in the *Surya Siddhanta*, a series of Hindu astronomy texts from about the year 400, and were developed by Aryabhata in *Aryabhatiya* around 500. Similar ideas evolved independently in China. The Indian tradition was continued by Varahamihira, Brahmagupta, and Bhaskaracharya, whose works include useful approximations to the sine function and some of the basic formulas, such as Varahamihira's

$$\sin^2\theta + \cos^2\theta = 1$$

which is the trigonometric interpretation of Pythagoras's Theorem.

Until recently scholars thought that Indian mathematics stagnated after Bhaskaracharya, followed only by commentaries rehashing the classical results. Only when Britain added India to its burgeoning empire did new mathematics appear there. This may have been true for large parts of India, but not for Kerala. Joseph[2] remarks that 'the quality of the mathematics available from the [Kerala school] texts ... is of such a high level compared to what was produced in the classical period that it seems impossible for the one to have sprung from the other.' However, the only comparable ideas are those developed centuries later in Europe, so no plausible 'missing link' is evident. The Kerala school's advances seem to have been *sui generis*.

Jyesthadeva's commentary *Yuktibhasa* describes a series attributed to Madhava:

> The first term is the product of the given sine and radius of the
> desired arc divided by the cosine of the arc. The succeeding terms
> are obtained by a process of iteration when the first term is
> repeatedly multiplied by the square of the sine and divided by the
> square of the cosine. All the terms are then divided by the odd
> numbers 1, 3, 5, ... The arc is obtained by adding and subtracting
> respectively the terms of odd rank and those of even rank.

Translated into modern notation, and remembering that the tangent $\tan \theta$ is the sine divided by the cosine, this becomes

$$\theta = \tan \theta - \frac{1}{3}\tan^3\theta + \frac{1}{5}\tan^5\theta - \frac{1}{7}\tan^7\theta + \ldots$$

Which (rewritten in terms of the inverse tangent) is what we in the west call Gregory's series – discovered in *our* civilisation by James Gregory in 1671, or perhaps a little earlier. According to the *Mahajyanayana Prakara* (Methods for the Great Sines), Madhava used this series to calculate π. A special case ($\theta = \pi/4 = 45°$) of the previous series gives an infinite series for π, the first example of its type:

$$\frac{\pi}{4} = 1 - \frac{1}{3} + \frac{1}{5} - \frac{1}{7} + \frac{1}{9} - \frac{1}{11} + \ldots$$

This isn't a practical way to calculate π, because the terms decrease very slowly and huge numbers of terms are needed for even a few decimal places. Taking $\theta = \pi/6 = 30°$ instead, Madhava derived a variant that converges faster:

$$\pi = \sqrt{12}\left(1 - \frac{1}{3 \times 3} + \frac{1}{5 \times 3^2} - \frac{1}{7 \times 3^3} + \ldots\right)$$

He calculated the first 21 terms to obtain π to 11 decimal places. This series was the first new method for computing π after Archimedes's use of ever finer polygons.

One aspect of Madhava's works is surprisingly sophisticated. He estimated the error when the series is truncated at some finite stage. In fact, he stated three expressions for the error, which can be added in as a correction term to improve the accuracy. His expressions for the error after adding n terms of the series are:

$$\frac{1}{4n} \qquad \frac{n}{4n^2 + 1} \qquad \frac{n^2 + 1}{4n^3 + 5n}$$

He used the third one to obtain an improved value for the sum, getting 13 decimal places of π. Nothing similar occurs anywhere in the mathematical literature until modern times.

In 1676 Newton wrote a letter to Henry Oldenburg, Secretary of the Royal Society, informing him of two infinite series for the sine and cosine:

$$\sin\theta = \theta - \frac{\theta^3}{3!} + \frac{\theta^5}{5!} - \frac{\theta^7}{7!} + \frac{\theta^9}{9!} - \frac{\theta^{11}}{11!} + \ldots$$

$$\cos\theta = 1 - \frac{\theta^2}{2!} + \frac{\theta^4}{4!} - \frac{\theta^6}{6!} + \frac{\theta^8}{8!} - \frac{\theta^{10}}{10!} + \ldots$$

which he had derived by a roundabout method, using calculus. We now know that these expressions, long assumed to have originated with Newton, were obtained by Madhava, nearly four centuries earlier. Details of the derivation of these series are given in *Yuktibhasa*. The method is complicated, but can be viewed as an early anticipation of the calculus method of integrating a series term by term.

Indeed, it is argued that Madhava developed some basic notions of calculus, long before Newton. Namely differentiation, the integral as the area under a curve, and term-by-term integration. He found methods for expanding polynomials in algebra, devised a numerical method for solving equations by iteration, and worked on infinite continued fractions.

Joseph asks whether Madhava's ideas might have percolated into Europe. He points out that European explorers such as Vasco da Gama knew Kerala well, because it is a useful stopping point for ships crossing the Arabian Sea *en route* to China and other places in the Far East. Its role as a trade centre goes back to Babylonian times. Its geographical isolation, hemmed in by the Western Ghats and the Arabian Sea, protected it from the turbulent politics of the rest of medieval India, an added bonus for foreign travellers. It does seem that some of Kerala's technology, and its goods, made their way to Europe at that time, but so far no evidence of direct transfer of mathematical ideas has been found. Until and unless new evidence comes to light, it appears that Kerala and Europe discovered many important mathematical ideas independently.

The work of great Indian figures like Aryabhata and Brahmagupta has long been recognised in Europe. That of the Kerala school was first brought to European scholarly attention as recently as 1835, when Charles Whish wrote an article about four major texts: Nilakantha's *Tantrasamgraha*, Jyesthadeva's *Yuktibhasa*, Putumana Somayaji's *Karana Paddhati*, and Sankara Varman's *Sadratnamala*. Whish put the cat among the pigeons with the claim that *Tantrasamgraha* contains the basis of fluxions, Newton's term for calculus (Chapter 7): that it 'abounds with fluxional forms and series to be found in no work of foreign countries'. In the days when the East India Company controlled trade with India, and the country itself was seen as ripe for conquest, this claim went down like a lead balloon. Kerala mathematics was essentially forgotten. Over a century later, in the 1940s, its advanced nature finally re-emerged in a series of articles by Cadambur Rajagopal and collaborators, analysing Kerala mathematics and demonstrating that Hindu mathematicians discovered many important results much earlier than the Europeans to whom they had generally been credited.

5

The Gambling Astrologer
Girolamo Cardano

Girolamo (Gerolamo, Geronimo) Cardano / Hieronymus Cardanus
Born: Pavia, Duchy of Milan, 24 September 1501
Died: Rome, 21 September 1576

AT A VERY EARLY PERIOD in my life, I began to apply myself seriously to the practice of swordsmanship of every class, until, by persistent training, I had acquired some standing even among the most daring. By night, even contrary to the decrees of the Duke, I armed myself and went prowling about the cities in which I dwelt. I wore a black woollen hood to conceal my features, and put on shoes of sheep-pelt. Often I wandered abroad throughout the night until day broke, dripping with perspiration from the exertion of serenading on my musical instruments.

Such was life in Renaissance Italy around 1520 – at least, for Girolamo Cardano, who revealed these activities and much more in a frank autobiography *The Book of My Life*. Cardano, a polymath especially gifted in mathematics and medicine, enjoyed (if that is the word) a career straight out of the soaps and tabloids. He frittered away the family fortune, became addicted to gambling, and endured ruin and the poor house. Suspecting another man was cheating, he slashed the player's face with a knife. He was accused of heresy and imprisoned; his son was executed for wife-poisoning. But Cardano also restored the speech of the Bishop of St Andrews, who had become mute, earning a reward of 1400 gold crowns. Returning to Italy in triumph, he was admitted into the College of Physicians, which had spent decades trying desperately to keep him out.

Most importantly of all, he was a master mathematician who wrote one of the all-time great textbooks, *Ars Magna* (The Great Art). Its subtitle: *The Rules of Algebra*. In *Ars Magna*, algebra came of age, acquiring both symbolic expression and systematic development. Cardano can be viewed as yet another candidate for the title 'father of algebra'. But, true to form, he did not attain this status without controversy and double-dealing.

Cardano was an illegitimate child. His father Fazio, a lawyer with strong mathematical talents and a hair-trigger temper, lived in Pavia and was a friend of Leonardo da Vinci. He habitually wore an unusual purple cloak and a small black skullcap, and had lost all his teeth by the age of 55. Girolamo's mother Chiara (*née* Micheria), a young widow with three children, married his father many years later. She was fat, with a temper to rival Fazio's, and quick to take offence. She was also deeply religious and highly intelligent. When she became pregnant, plague appeared in Milan, so she moved to the countryside, while her three elder children remained in the city and died of the plague. Cardano's imminent arrival did not provoke joy: 'Although various abortive medicines ... were tried in vain, I was normally born on the 24th day of September in the year 1500.'

Fazio, though a lawyer by trade, was sufficiently adept at mathematics to have advised da Vinci about geometry, and he taught geometry at the University of Pavia and at the Piatti Foundation in Milan. He passed on his skill in mathematics and astrology to his bastard son: 'My father, in my earliest childhood, taught me the rudiments of arithmetic, and about that time made me acquainted with the arcane ... He instructed me in the elements of the astrology of Arabia ... After I was twelve years old he taught me the first six books of Euclid.'

Girolamo was a sickly child, and his father's plans to bring him into the family's legal business failed. Enrolling as a medical student at Pavia University, he performed brilliantly, and although many found his outspoken nature offensive, he was elected rector of the university by the margin of a single vote. Success went to his head. This was the period when he roamed the city streets armed with his sword and musical instruments, and turned to gambling. His mathematical understanding of chance gave him a distinct advantage, and around 1564 he wrote one of the earliest books on probability: *Book on Games of Chance*, finally published in 1663. His ability to play chess – for money – also helped. But as he became more dissolute, he lost both his luck and his inheritance.

Still he pressed on. Now in possession of a medical degree, he tried to gain entry to Milan's College of Physicians, the gateway to a profitable profession and a comfortable lifestyle. This time his tendency to speak his mind frankly let him down, and he was refused entry, so he took a position as a doctor in a nearby village. It brought in just about enough to live on, and he married Lucia Bandarini, daughter of a militia captain. Rejected again by the college, he reverted to his earlier occupation, and lost a fortune. After he'd pawned all their possessions, including Lucia's jewellery, they ended up in the poor house. 'I ruined myself! I perished!' Cardano wrote. He and Lucia had a child, who suffered from various minor birth defects but did not, at that time, count as deformed. By now Fazio had died, and Girolamo was appointed his successor; things were finally looking up. In 1539 even the College of Physicians stopped trying to keep him out. He was developing a new string to his bow as well, publishing several mathematics books. One of them placed him firmly in the ranks of the mathematical trailblazers.

✳

Most areas of mathematics emerge through a complex and confusing historical process that lacks any clear direction, precisely because the direction itself is being created as fragmentary ideas begin to link together. The jungle grows as you explore it. A few features of algebra can be traced back to the ancient Greeks, who lacked an effective notation even for whole numbers. By inventing abbreviated notation for unknown quantities, Diophantus gave proto-algebra a big boost, but his focus was on solving equations in whole numbers, which led more directly to number theory. Greek and Persian geometers solved problems we now consider to be algebraic by purely geometric means. Al-Khwarizmi formalised algebraic processes, but failed to introduce symbolic representations.

Long before any of this had happened, the Babylonians had already discovered the first genuinely important technique in algebra: how to solve quadratic equations. This kind of question, we now appreciate, opens the door to algebra in the form it had acquired by the nineteenth century, which is the bulk of what goes by that name in school mathematics. Namely, deducing the value (or a short list of possible values) of an unknown quantity from some numerical relationship between that quantity and its 'powers' – its square, its cube, and so on. That is, solving a polynomial equation.

If the highest power of the unknown that appears is its square, the equation is quadratic. The scribe-mathematicians of ancient Babylon knew how to solve such things, and they taught them to schoolboys. We have the clay tablets, with their arcane cuneiform characters, to prove it. The most difficult step is taking the square root of an appropriate quantity.

With hindsight, the next step is clear: cubic equations, involving the cube of the unknown as well as its square and the unknown itself. One Babylonian tablet hints at a special method for solving cubics (the nickname that mathematicians use in place of the cumbersome 'cubic equations'), but that's all we know of their discoveries in this area. The Greek and Persian geometric methods did the trick; the most detailed treatment is that of Omar Khayyam, more famous for his poetry, especially *Rubaiyat* (and even more especially in Edward FitzGerald's translation). A purely algebraic solution seemed beyond reach.

All that changed in the heady days of the Italian Renaissance.

Around 1515 Scipione del Ferro, a professor in Bologna, discovered how to solve some types of cubic. The distinction into types arose because negative numbers were not then recognised, so equations were arranged with positive terms on both sides. Del Ferro passed on some notes to his son-in-law Annibale del Nave, which show that he could solve the case 'cube plus unknown equals number'. In all likelihood he could also solve two other types, which between them effectively cover all possibilities after some preparatory manipulation. His method involved both square roots and cube roots.

Along with del Nave, the method for the aforementioned case was known to del Ferro's student Antonio Fior. Independently, Niccolò Fontana (generally known by the now politically incorrect nickname of Tartaglia, 'stammerer') rediscovered the solution for the same case. Fior, who was intending to set up shop as a mathematics teacher, got a bright idea: engage Tartaglia in a public battle, where each would challenge the other to solve mathematical problems. This sort of intellectual combat was common at the time. But the scam blew up in his face when Tartaglia, spurred on by rumours that all three cases had been solved, and deeply worried that Fior knew how, made a huge effort and found the solutions just in time for the contest. Belatedly discovering that Fior could solve only one type, Tartaglia then set him only cases that he couldn't solve, and wiped the floor with him.

The news was juicy and spread rapidly, reaching the ears of Cardano, who was assiduously collecting material for *Ars Magna*. He, keeping a weather eye open for any interesting new mathematics that might improve his intended book, was quick to spot a golden opportunity. Del Ferro's earlier work had by then been all but forgotten, so Cardano visited Tartaglia, begging him to reveal the secret of the cubic. Eventually Tartaglia gave in. Legend has it that he swore Cardano to secrecy, but this seems a little unlikely in the context of Cardano's intention to write an algebra book. At any rate, when *Ars Magna* appeared, it contained Tartaglia's solution of the cubic. Credited to him, but that came a poor second to being scooped. The irate Tartaglia hit back with *Diverse Questions and Inventions*, which included all of the correspondence between him and Cardano. The book claimed that in 1539 Cardano had sworn a solemn oath 'never to publish your discoveries'. Now, the oath was broken.

As might be expected, the full story is probably more complicated. Some time afterwards, Lodovico Ferrari, who later became Cardano's student, claimed he'd been present at the meeting, and Cardano had not agreed to keep Tartaglia's method secret. On the other hand, Ferrari was hardly a dispassionate observer. His response to Tartaglia's claim of a broken oath was to issue a *cartello* – challenging Tartaglia to debate with him on any topic he wished. In August 1548 a large crowd gathered in a church to watch the contest. I doubt many were drawn there by the mathematics, or even understood it: what most of them wanted was a good old set-to. Although no records of the outcome are known, Ferrari was soon offered the post of tutor to the Emperor's son; in contrast, Tartaglia never claimed a win, lost his job in Brescia, and kept whingeing about the result. So we can make an educated guess.

The irony is that none of it had been necessary. During the preparation of *Ars Magna*, Cardano and Ferrari had seen del Ferro's Bologna papers, which contained his prior solution of the cubic. This, they maintained, had been the actual source of the method. The reason Cardano had mentioned Tartaglia's work had been to explain how he had heard of del Ferro's. That was all.

Perhaps. But why did Cardano beg Tartaglia to reveal the secret, if he already knew it from an earlier source? Maybe he didn't beg. We have only Tartaglia's word for it. On the other hand, *something* held Cardano back, for a time, because he didn't just need the solution of the cubic in its own right. Ferrari, under Cardano's guidance, had managed to take everything one step further by solving the quartic equation (fourth power of the unknown, as well as lower powers). But, crucially, his solution operated by reducing everything to a related cubic. So Cardano couldn't reveal to the world the solution of the quartic, without also telling them how to solve cubics.

Perhaps it was all as Cardano and Ferrari claimed. Tartaglia's defeat of Fior made Cardano aware that cubics could be solved.

Then a bit of digging led him to del Ferro's manuscript, which gave him the method he needed for his book. Stimulated by this discovery, Ferrari then conquered the quartic. Cardano put the lot in his book. Ferrari, as his student, could hardly complain about his results being included, and seems to have taken pride that they were. Out of deference to Tartaglia, Cardano gave him credit for rediscovering the method and drawing it to his attention.

Ars Magna is significant for one other reason. Cardano applied his algebraic methods to find two numbers whose sum is 10 and product is 40, and got the answer $5 + \sqrt{-15}$ and $5 - \sqrt{-15}$. Since negative numbers have no square roots, he declared this result to be 'as subtle as it is useless'. The formula for cubics also leads to such quantities when all three solutions are real, and in 1572 Rafael Bombelli observed that if you ignore what such expressions might mean and just do the sums, you get the correct real solutions. Eventually this line of thinking led to the creation of the system of complex numbers, in which -1 has a square root. This extension of the real number system is essential to today's mathematics, physics, and engineering.

In the 1540s Cardano went back to practising medicine. Then (like I said, soaps and tabloids) tragedy struck. His eldest son Giambatista had secretly married Brandonia di Seroni, who in Cardano's opinion was worthless and shameless. Her parents were gold-diggers, and Giambatista's wife taunted him in public, claiming that he wasn't the father of their three children. He poisoned her and promptly confessed. The judge insisted that the only way to avoid the death penalty was if Cardano agreed compensation with the di Seronis. The sum they demanded was so huge that he couldn't pay, so his son was tortured, had his left hand cut off, and was then beheaded.

Cardano, a tough cookie who'd seen it all, was forced to move, becoming professor of medicine at Bologna. There, his arrogance made enemies of his medical colleagues, and they tried to get him fired. His younger son Aldo amassed huge gambling debts, and broke into his father's house to steal money and jewels. Cardano felt he had no choice but to report the theft, and Aldo was banished from Bologna. Even so, Cardano remained an optimist, writing that despite these tragic events, 'I still have so many blessings, that if they were another's he would count himself lucky.' But more disasters were in the pipeline of fate, and their cause was his involvement in astrology. In 1570 he cast Jesus's horoscope. He also wrote a book praising Nero, who had martyred early Christians. The combination led to a charge of heresy. He was imprisoned, and then released, but barred from any academic post.

He went to Rome, where to his surprise he was greeted warmly. Pope Gregory XIII had apparently forgiven him, and granted him a pension. He was admitted to the College of Physicians there, and wrote, but did not publish, his autobiography. It finally appeared more than sixty years after his death. According to legend, he died by his own hand because he had predicted his own date of death, and professional pride required the forecast to be correct.

6

The Last Theorem
Pierre de Fermat

Pierre de Fermat
Born: Beaumont-de-Lomagne, France, 17 August 1601
(or 31 October – 6 December 1607)
Died: Castres, France, 12 January 1665

FEW MATHEMATICIANS MANAGE to pose a problem that remains unanswered for centuries, especially one that turns out to be of central importance to areas of the subject that didn't even exist when the question was first asked. Pierre Fermat (the 'de' was added later when he became a government official) is perhaps the best known among this exalted company. But he wasn't exactly a mathematician; he had a law degree and became a councillor at the parliament in Toulouse. On the other hand, it would be stretching a point to call him an amateur. Perhaps he is best thought of as an unpaid professional whose day job was practising law.

Fermat published very little, possibly because his non-mathematical duties left hardly any time to write up his discoveries. What we know of them comes mainly from his letters to mathematicians and philosophers, such as Pierre de Carcavi, René Descartes, Marin Mersenne, and Blaise Pascal. Fermat knew what a proof was, and in particular the sole incorrect statement in his surviving papers (about a formula that he thought always yielded a prime number) was accompanied by the assertion that he lacked a proof. Very few of his proofs survive, the main one being a proof that two squares can't add up to a fourth power, accomplished by a novel method that he called 'infinite descent'.

Fermat has many claims to mathematical fame. He made major advances in geometry, developed precursors to calculus, and worked on probability and the mathematical physics of light. His foremost contribution, however, is his seminal work in number theory. There, he stated the conjecture that also ensured his fame among the general public, thanks in part to a television documentary and a bestselling book. Namely, his Last Theorem. This simple but enigmatic statement acquired its name not because he gasped it out on his deathbed, but because Fermat's successors, over the next hundred years or so, managed to prove (or in one case disprove) every theorem he had stated, with this sole exception. It was the last to hold out against the onslaught, and it baffled the finest minds.

Among them was Gauss, one of the finest of them all. Nearly two hundred years after Fermat's marginal note, Gauss dismissed Fermat's Last Theorem, declaring it typical of a huge range of statements about numbers that are easy to guess but virtually impossible to prove or disprove. Gauss normally had impeccable taste when it came to mathematics, but this assessment turned out to be a rare underestimate of mathematical significance. In Gauss's defence, most mathematicians felt the same way for the first three and a quarter centuries after Fermat stated the problem. Only when subtle links to other, more central areas of mathematics, were spotted, did its true importance emerge.

Today, Beaumont-de-Lomagne is a French commune (an administrative district) in the Midi-Pyrénées region of southern France. It was founded in 1276 as a bastide – one of a series of fortified medieval towns in that area, and it had a turbulent history. The town was captured temporarily by the English during the Hundred Years War and then lost 500 of its citizens to the plague. It was Catholic, hemmed in by three neighbouring Protestant towns. Henri III sold it to the future Henri IV, who attacked it in 1580 and massacred a hundred of its people. Louis XIII put it under siege in the early 1600s; it took part in the rebellion against the king and was subject to military occupation in 1651 and heavily fined; then plague struck again.

Neatly bookended by these events was the birth of the town's most famous inhabitant: Pierre Fermat, son of a rich leather merchant Dominique and his wife Claire (*née* de Long) who hailed from a family of lawyers. There's some uncertainty about the year of his birth (either 1601 or 1607) because he might have had an elder brother, also named Pierre, who died young. His father was also the second consul of Beaumont-de-Lomagne, so Fermat grew up in a political family. His father's position makes it a racing certainty that Fermat grew up in the town of his birth, and if so, he must have been educated in its Franciscan monastery. After a spell at the University of Toulouse he went to Bordeaux, where his mathematical interests began to flourish. First he produced a tentative restoration of *On Plane Loci*, a lost work of the Greek geometer Apollonius; then he wrote on maxima and minima, anticipating some early developments in calculus.

His legal career also blossomed with a law degree from the University of Orléans. In 1631 he bought himself a position as a councillor in the parliament at Toulouse, which entitled him to add 'de' to his name. He acted in this capacity, and as a lawyer, for the rest of his life, living in Toulouse but working from time to time in Beaumont-de-Lomagne and Castres. Initially he was in the lower chamber of parliament, but he rose to a higher chamber in 1638 and thence, in 1652, to the top level of the criminal court. Helped in part by the plague, which killed many senior officials in the 1650s, Fermat continued to be promoted. In 1653 it was reported that Fermat had died of the plague, but (as with Mark Twain) the report was greatly exaggerated. It looks as though Fermat was biting off more than he could chew; his interest in mathematics was distracting him from legal duties. One document reads: 'He is rather preoccupied, he does not report cases well and is confused.'

His 1629 *Introduction to Plane and Solid Loci* pioneered the use of coordinates to link geometry and algebra. This idea is generally credited to Descartes in his 1637 essay *La Geometrie*, an appendix to *Discours de la Méthode*, but it was hinted at in much earlier writings, all the way back to the Greeks. It uses a pair of coordinate axes to represent each point of the plane by a unique pair of numbers (x, y), a method now so commonplace that it scarcely requires discussion.

In his 1679 *On Tangents to Curved Lines*, Fermat found tangents to curves, a geometric version of differential calculus. His method for finding maxima and minima was another forerunner of calculus. In optics he stated the principle of least time: a light ray follows whichever path minimises the total travel time. This was an early step towards the calculus of variations, a branch of analysis that seeks curves or surfaces that minimise or maximise some related quantity. For example, which closed surface of fixed volume has the smallest surface area? The answer is a sphere, and this explains why soap bubbles are spherical, because the energy of surface tension is proportional to the surface area, and the bubble takes whichever shape makes that energy smallest.

In a similar vein, Fermat argued with Descartes over the latter's deduction of the law of refraction for light rays. Descartes, probably annoyed that Fermat was getting credit for coordinates, which Descartes considered his own invention, responded by criticising Fermat's work on maxima, minima, and tangents. The dispute became so heated that the engineer and pioneering geometer Girard Desargues was roped in as referee. When he said Fermat was right, Descartes grudgingly conceded: 'If you had explained it in this manner at the outset, I would have not contradicted it at all.'

＊

Fermat's greatest legacy is in number theory. His letters contain many challenges to other mathematicians. These included proving that the sum of two perfect cubes can't be a perfect cube, and solving the misnamed 'Pell equation' $nx^2 + 1 = y^2$, where n is a given whole number and whole numbers x and y must be found. Leonhard Euler erroneously attributed a solution by Lord Brouncker to John Pell. In fact, Brahmagupta's *Brahmasphutasiddhanta* (Correctly Established Doctrine of Brahma) of 628 includes a method for solving it.

One of Fermat's most important and beautiful results characterises those numbers that can be expressed as the sum of two perfect squares. Albert Girard stated the answer in work published posthumously in 1634. Fermat was the first to claim a proof, announcing it in a letter to Mersenne in 1640. The main point is to solve the problem for prime numbers. The answer depends on the type of prime, in the following sense. The only even prime is 2. Odd numbers are either a multiple of 4 with 1 added, or a multiple of 4 with 3 added; that is, they are of the form $4k + 1$ or $4k + 3$. The same goes for odd primes, of course. Fermat proved that 2, and every prime of the form $4k + 1$, are sums of two squares; on the other hand, those of the form $4k + 3$ are not.

If you experiment, it's easy to guess this. For example $13 = 4 + 9 = 2^2 + 3^2$, and $13 = 4 \times 3 + 1$. On the other hand $7 = 4 \times 1 + 3$, and a sum of two squares can't equal 7. Proving Fermat's two-squares theorem, though, is distinctly hard. The easiest bit is to show that the $4k + 3$ primes are not sums of two squares; I'll show you how in Chapter 10 using a trick that Gauss developed to systematise a basic method of number theory. Showing that the $4k + 1$ primes *are* sums of two squares is considerably harder. Fermat's proof hasn't survived, but proofs are known that use methods that would have been accessible to him. Euler gave the first known proof, announcing it in 1747 and publishing it in two articles of 1752 and 1755.

The upshot is that a whole number is the sum of two squares if and only if every prime factor of the form $4k + 3$ appears to an even power when the number is resolved into prime factors. For instance,

$245 = 5 \times 7^2$. The factor 7 is of the form $4k + 3$, and occurs to an even power, so 245 is the sum of two squares. Indeed, $245 = 14^2 + 7^2$. In contrast, $35 = 5 \times 7$, and the factor 7 occurs to an odd power, so 35 is not the sum of two squares. This result may seem an isolated curiosity, but it sparked several lines of research, flowering into Gauss's far-reaching theory of quadratic forms (Chapter 10). In modern times it has been taken much further. A related theorem, proved by Lagrange, states that any whole number is the sum of four squares (where $0 = 0^2$ is allowed). This, too, has extensive ramifications.

The story of Fermat's Last Theorem has been told many times, but I make no apology for telling it again. It's a great story.

It's perhaps ironic that Fermat's greatest fame rests on a theorem that he almost certainly didn't prove. He apparently *claimed* a proof, and the result is now known to be true, but the verdict of history is that the methods available to him weren't up to the task. His claim to possess a proof exists only as a marginal note in a book, which doesn't even survive as an original document, so it could have been made prematurely. In mathematical research it's not unusual to wake up in the morning convinced you've proved something important, only to see the proof evaporate by noon when you find a mistake.

The book concerned was a French translation of the *Arithmetica* of Diophantus, the first great work on number theory, unless you count Euclid's *Elements*, which develops many basic properties of prime numbers and solves some important equations. Certainly *Arithmetica* is the first specialist text on the topic. Recall that this book gave mathematics the technical term 'Diophantine equation' for a polynomial equation that must be solved in whole or rational numbers. Diophantus made a systematic catalogue of such equations, and one of the central exhibits is the equation $x^2 + y^2 = z^2$ for so-called Pythagorean triples, because a triangle with sides x, y, and z has a right angle by Pythagoras's Theorem. The simplest solution in nonzero whole numbers is $3^2 + 4^2 = 5^2$, the celebrated 3–4–5 triangle. There are infinitely many solutions: Euclid gave a procedure to generate all of them; Diophantus included it.

Fermat owned a copy of Claude Bachet de Méziriac's 1621 Latin translation of *Arithmetica*, and he jotted observations in the margins. According to Fermat's son Samuel, the Last Theorem is stated as a note attached to Diophantus's Question VIII of Book II. We know this because Samuel issued his own edition of *Arithmetica*, which included his father's notes. The dates of the notes are not known, but Fermat started studying *Arithmetica* around 1630. The date is often given as 1637, but this is a guesstimate. Presumably musing on potential generalisations of Pythagorean triangles, Fermat was led to his epic marginal annotation:

> It is impossible to divide a cube into two cubes, or a fourth power into two fourth powers, or in general, any power higher than the second, into two like powers. I have discovered a truly marvellous proof of this, which this margin is too narrow to contain.

That is, the Diophantine equation $x^n + y^n = z^n$ has no whole number solutions if n is an integer greater than or equal to three.

There's circumstantial evidence that Fermat subsequently changed his mind about having a proof. In correspondence, he often set his theorems as puzzles for other mathematicians to solve (and at least one of them complained they were too hard). However, none of his extant letters mentions this theorem. Even more tellingly, he did pose two special cases, cubes and fourth powers, as problems for his correspondents. Why do this if he could prove a more general result? It seems certain that he could prove the cubic case, and we know how he proved the theorem for fourth powers. In fact, that proof is the *only* proof in all of the works and papers he left. His actual statement is: 'The area of a right triangle cannot be a square.' He clearly intended this to refer to Pythagorean triples. The Euclid–Diophantus solution easily implies that this problem is equivalent to finding two squares that sum to a fourth power. If a solution of $x^4 + y^4 = z^4$ with exponent 4 existed, then both x^4 and y^4 would be squares (of x^2 and y^2 respectively); Fermat's statement then implies that no such solution exists.

His proof is ingenious, and at the time was a radical innovation. He called it the method of infinite descent. Suppose a solution exists, apply the Euclid–Diophantus solution, mess around for a bit, and you can deduce that a *smaller* solution exists. Therefore, said

Fermat, you can construct an infinite chain of ever smaller solutions. Since any descending chain of this kind, formed from whole numbers, must eventually stop, this is a logical contradiction. So the hypothetical solution that we started from can't actually exist.

Fermat may have concealed his proofs deliberately. He seems to have been rather mischievous, and liked to torment other mathematicians by setting his results as puzzles. His marginal note is not the only one to announce something important followed immediately by an excuse for not proving it. Descartes considered Fermat to be a braggart, and Wallis referred to him as 'that damned Frenchman'. Be that as it may, the tactic – if such it was – worked. After Fermat's death – indeed, during his life too – great mathematicians made their mark by polishing off one or other of the puzzles he had posed to posterity. Euler, for instance, claimed a proof that two cubes can't add to a cube in a letter of 1753 to his friend Christian Goldbach. We now realise that his proof had a gap, but this can be fixed up fairly easily, so Euler generally gets credit for the first published proof. Adrien-Marie Legendre proved the Last Theorem for fifth powers in 1825, and Peter Dirichlet proved it for 14th powers in 1832, clearly a failed attempt to prove it for seventh powers that could be salvaged by aiming for something weaker. Gabriel Lamé dealt with seventh powers in 1839, and in 1847 he explained the main ideas of the proof to the Paris Academy of Sciences. It involved an analogue of prime factorisation for a special type of complex number.

Immediately after his talk, Joseph Liouville stood up and pointed out a possible flaw in Lamé's method. For the usual kind of number, prime factorisation is *unique*: aside from the order in which the factors are written, there's only one way to do it. For example, the prime factorisation of 60 is $2^2 \times 3 \times 5$ and nothing essentially different works. Liouville was worried that unique factorisation might not be valid for Lamé's class of complex numbers. Eventually his doubts were justified: the property first fails for 23rd powers.

Ernst Kummer managed to fix up this idea by throwing new ingredients into the mix, which he called 'ideal numbers'. These behave like numbers, but aren't. He used ideal numbers to prove Fermat's Last Theorem for many powers, including all primes up to 100 except 37, 59, and 67. By 1993 Fermat' s Last Theorem was known to be true for all powers up to 4 million, but this kind of increasingly desperate scramble wasn't shedding any light on the general case. New ideas started to show up in 1955 when Yutaka Taniyama was working on a different, apparently unrelated area of number theory called elliptic curves. (The name is misleading and the ellipse isn't one of them. An elliptic curve is a special kind of Diophantine equation.) He conjectured a remarkable link between these curves and complex analysis, the theory of modular functions. For years hardly anyone believed he was right, but evidence slowly piled up that what is usually called the Shimura–Taniyama–Weil conjecture might actually be true.

In 1975 Yves Hellegouarch noticed a relationship between Fermat's Last Theorem and elliptic curves, suggesting that any counterexample to Fermat's Last Theorem would lead to an elliptic curve with very strange properties. In two papers published in 1982 and 1986, Gerhard Frey showed that this curve must be so strange that it can't exist. Fermat's Last Theorem would then follow by contradiction, except that Frey made essential use of the Shimura–Taniyama–Weil conjecture, which was still up for grabs. However, these developments convinced many number theorists that Hellegouarch and Frey were on the right track. Jean-Pierre Serre predicted that someone would prove Fermat's Last Theorem by this route, about a decade before it happened.

Andrew Wiles took the final step in 1993, announcing the proof of a special case of the Shimura–Taniyama–Weil conjecture, powerful enough to complete the proof of Fermat's Last Theorem. Unfortunately a logical gap then came to light, often a prelude to total collapse. Wiles was lucky. Helped by his former student Richard Taylor, he managed to repair the gap in 1995. Now the proof was complete.

People still debate whether Fermat had a proof. As I've said, circumstantial evidence strongly suggests he didn't, because he would surely have posed it as a challenge to others. More likely, he thought he had a proof when he scribbled his marginal note, but later changed his mind. In the unlikely event that he did have a proof, it can't have been anything like Wiles's. The necessary concepts, and the abstract viewpoint, simply did not exist in Fermat's day. It's like expecting Newton to have invented nuclear weapons. Still, it's conceivable that Fermat spotted some method of attack that everyone else had missed. Such things have happened. However, no one is going to find such a proof unless they have the mathematical talents of Pierre de Fermat, and that's a tall order.

7

System of the World
Isaac Newton

Sir Isaac Newton
Born: Woolsthorpe, England, 4 January 1643
Died: London, 31 March 1727

IN 1696 THE ROYAL MINT, responsible for the production of England's money, acquired a new warden, Isaac Newton. He had been granted the position by the Earl of Halifax, Charles Montagu, who at that time was Chancellor of the Exchequer – the government's head of finance. Newton was put in charge of the recoining of the realm. At that time, Britain's coinage was in a shoddy state. Newton estimated that about 20 per cent of the coins in circulation were either counterfeit or clipped; that is, slivers of gold or silver had been shaved off their edges, to be melted down and sold. In principle, counterfeiting and clipping were acts of treason, punishable by the judicial torture of being hanged, drawn, and quartered. In practice, hardly anyone was ever convicted, let alone punished.

As Lucasian Professor of Mathematics at Cambridge University, the new warden was an ivory-tower academic who had devoted most of his life to the esoteric subjects of mathematics, physics, and alchemy. He had also written religious tracts about interpretations of the Bible, and had dated the Creation to 4000 BC. His track record of public service was patchy. He had served as Member of Parliament for Cambridge University in 1689–90 and would do so again in 1701–2, but it's claimed that his sole contribution to debates was that he once observed that the chamber was cold and asked for the windows to be closed. So it was easy to assume that, having obtained the post as a sinecure through political patronage, Newton would be a pushover.

Within a few years, 28 convicted coiners would realise that this was not the case. Newton went in search of evidence in a manner that would do credit to Sherlock Holmes. He disguised himself as a frequenter of disreputable taverns and alehouses, spying on the customers, observing criminal activity. Realising that the obfuscatory nature of English law was one of the biggest obstacles to a successful prosecution, Newton fell back on the country's ancient customs and legal precedents. The office of justice of the peace had considerable legal authority, able to open prosecutions, cross-examine witnesses, and pretty much act as judge and jury. So Newton got himself appointed as a justice of the peace in all of the counties adjacent to London. In eighteen months, starting in the summer of 1698, he cross-examined more than a hundred witnesses, suspects, and informers, securing the aforementioned 28 convictions.

We know about this, incidentally, because Newton stuck a draft letter about it in his own copy of his masterwork *Principia*, in which he effectively founded mathematical physics by stating the laws of motion and the law of gravity, and showing how they explain a vast range of natural phenomena.

The tale illustrates that when Newton turned his mind to something, he usually achieved great things, though not in alchemy and probably not in biblical scholarship. He went on to become Master of the Mint, President of the Royal Society, and was knighted by Queen Anne in 1705. His greatest contributions to humanity, however, were in mathematics and physics. He invented calculus and used it to express fundamental laws of nature, from which he deduced – as the subtitle of Book 3 of the *Principia* states – the System of the World. How the universe works.

His beginnings, however, were far humbler.

Newton was born on Christmas Day 1642. At least, that was the date when he was born. But it was determined by the Julian calendar, and when that was replaced by the Gregorian calendar, notorious for its 'lost days', the official date became 4 January 1643. As a child, he lived on a farm; Woolsthorpe Manor in the tiny village of Woolsthorpe-by-Colsterworth in Lincolnshire, not far from Grantham.

Newton's father, also named Isaac, died two months before his son was born. The Newtons were an established farming family, and Isaac Newton the elder was comfortably wealthy, owning a large farm, a house, and many farm animals. His mother Hannah (*née* Ayscough) managed the farm. When Isaac was two she remarried to Barnabas Smith, minister of the church in the nearby village of North Witham. The boy was cared for at Woolsthorpe by his grandmother Margery Ayscough. It wasn't a happy childhood, and Isaac didn't get on terribly well with his grandfather James Ayscough. He got on even worse with his mother and stepfather: when he confessed his sins at the age of 19, he mentioned 'threatening my father and mother Smith to burn them and the house over them'.

His stepfather died in 1653, and Isaac started at the Free Grammar School in Grantham, where he lodged with the Clarke family. William Clarke was an apothecary, and his house was on the High Street next to the George Inn. Newton became notorious among the townsfolk for his strange inventions and mechanical devices. He spent his pocket money on tools, and instead of playing games, he made things from wood – dolls' houses for the girls, but also a working model of a windmill. He added a treadmill, run by a mouse, to drive the mill. He made a small cart which he could sit in and move by turning a crank. And he fixed a paper lantern to a kite to startle the neighbours at night. According to his biographer William Stukeley, this 'wonderfully affrighted all the neighboring inhabitants for some time, and caus'd not a little discourse on market days, among the country people, when over their mugs of ale.'

Historians have since discovered Newton's source for most of these inventions, *The Mysteries of Nature and Art* by John Bate. One of Newton's notebooks contains numerous extracts from this book. But the inventions illustrate his early focus on scientific matters, even if they were not original. He was also fascinated by sundials – Colsterworth Church has one attributed to him, supposedly constructed when he was only nine – and young Isaac distributed sundials liberally throughout Clarke's house. He hammered pegs in the walls to mark the hours, half-hours, and quarter-hours. He learned how to spot significant dates, such as solstices and equinoxes, with enough success that the family and their neighbours often came to look at what they called 'Isaac's dials'. He could tell the time by observing the shadows in a room. He also took advantage of living in an apothecary's shop by investigating the composition of medicines, an early introduction to chemistry which paved the way to his extensive alchemical interests in later life. He drew impressively convincing birds, animals, ships, even portraits on the walls of his room in charcoal.

He was clearly a clever young man, but he showed no special signs of mathematical talent, and his school reports describe him as idle and inattentive. At this point his mother took him out of school to have him trained to manage her estate, a standard task for an eldest son, but he showed even less interest in that. An uncle persuaded her that Isaac should go to university in Cambridge, so she sent him back to Grantham to complete his education.

He entered Trinity College at Cambridge University in 1661, intending to obtain a degree in law. The course was based on Aristotelian philosophy, but in the third year he was allowed to read works by Descartes, the philosopher-scientist Pierre Gassendi, the philosopher Thomas Hobbes, and the physicist Robert Boyle. He studied Galileo, learning about astronomy and the Copernican theory that the Earth goes round the Sun. He read Kepler's *Optics*. How Newton was introduced to advanced mathematics is more opaque. Abraham de Moivre wrote that it all started when Newton bought an astrology book at a fairground, and couldn't understand the mathematics. Trying to master trigonometry, he discovered he didn't know enough geometry, so he picked up a copy of Isaac Barrow's edition of Euclid. That struck him as trivial, until he got to a theorem about the areas of parallelograms, which impressed him. He then raced through a series of major mathematical works – William Oughtred's *The Key of Mathematics*, Descartes's *La Géométrie*, the works of François Viète, Frans van Schooten's *Geometry of René Descartes*, and John Wallis's *Algebra*. Wallis used indivisibles – infinitesimals – to calculate the area contained by a parabola and a hyperbola. Newton thought about this, and wrote: 'Thus Wallis doth it, but it may be done thus ...' Already he was coming up with his own proofs and ideas, inspired by the great mathematicians, but not subservient to them. Wallis's methods were interesting, but by no means sacred. Newton could do better.

In 1663 Barrow took up the Lucasian chair, becoming a Fellow of Trinity, where Newton was based, but there's no evidence that he noticed any special talent in the young student. It came to flower in 1665, when the university's students were sent home to avoid the great plague. In the peace and quiet of the Lincolnshire countryside, no longer distracted by the hustle and bustle of the city, Newton turned his attention to science and mathematics. Between 1665 and 1666 he developed his law of gravity, which explained the movements of the Moon and planets, devised laws of mechanics to explain moving bodies, invented the differential and integral calculus, and made significant discoveries in optics. He published none of this work, but returned to Cambridge to take his master's degree, and was elected a Fellow of Trinity College. In 1669 he was appointed Lucasian Professor of Mathematics when Barrow resigned, and he became a Fellow of the Royal Society in 1672.

From 1690 Newton wrote many tracts on the interpretation of the Bible and carried out alchemical experiments. He held important administrative posts, eventually becoming Master of the Royal Mint. He was elected President of the Royal Society in 1703, and knighted in 1705 when Queen Anne paid a visit to Trinity College, Cambridge. The only scientist to have been knighted before then was Francis Bacon. He lost a fortune with the collapse of the South

Sea Bubble, and went to live with his niece and her husband near Winchester until dying in his sleep in London in 1727. Mercury poisoning has been suspected: traces of the metal were found in his hair. It fits with his experiments in alchemy, and could explain why he became eccentric in his old age.

One of Newton's early discoveries shows him to be a master of coordinate geometry. By then, conic sections were known to be defined by quadratic equations. Newton studied the curves defined by cubic equations. He found 72 species (we now recognise 78), and grouped them into four distinct types. In 1771 James Stirling proved that every cubic curve belongs to one of these types. Newton claimed that all four types are equivalent under projection, and a proof was found in 1731. In all of these discoveries, Newton was well ahead of his time, and the broad context into which they fit – algebraic and projective geometry – became apparent only centuries later.

According to a possibly apocryphal tale, one of Newton's practical inventions came into being during his early work on optics, around 1670. Every schoolchild is told that a glass prism splits white sunlight into all the colours of the rainbow. This discovery goes back to Newton, who performed the experiment in his attic. However, there was a snag. He had a cat, which apparently was rather rotund because its master, absorbed in his scientific research, failed to control how much it ate. The cat had a habit of pushing the attic door open to find out what Isaac was doing, which let light in and ruined the experiment. So Newton cut a hole in the door and hung a piece of felt over it, inventing the catflap. When the cat had kittens, he added a smaller hole next to the big one. (This may not have been as absurd as it appears; perhaps the kittens found it difficult to push past a heavy piece of felt.) The source for this anecdote has been identified only as a 'country parson', and it may just be a shaggy cat story. But in 1827 John Wright, who lived in Newton's former rooms at Trinity College, wrote that the door had once had two holes, of the right size for a cat and a kitten.

Newton's biggest contributions to mathematics, however, are calculus and the *Principia*. His work in optics made major strides in physics, but was less influential in mathematics, so I won't discuss it further. Logically, calculus comes before the *Principia*, but historically both are intertwined in complex ways, made all the more obscure by Newton's reluctance to publish. He had an instinctive dislike of criticism, and the easy way to avoid it was to keep his discoveries to himself. The end result, as it happened, was a much greater barrage of criticism and a huge public controversy, because the German mathematician and philosopher Gottfried Leibniz had very similar ideas around the same time, eventually triggering a priority dispute.

The origins of calculus can be traced back to Archimedes's *Method*, Wallis's 1656 *Arithmetic of the Infinite*, and works of Fermat (Chapter 6) The subject divides into two distinct but related areas.

Differential calculus is a method for finding the rate of change of some quantity that varies with time. For example, velocity is the rate of change of position (how many kilometres your position changes as an hour passes). Acceleration is the rate of change of velocity (are you speeding up or slowing down?). The basic issue in differential calculus is to find the rate of change of some function of time. The result is also a function of time, because the rate of change can be different at different times.

Integral calculus is about areas, volumes, and similar concepts. It proceeds by cutting the object into very thin slices, estimating the area or volume of each slice by ignoring any error that is much smaller than the thickness of the slices, adding everything together, and letting the slices become arbitrarily thin. As both Newton and Leibniz discovered, integration is essentially the reverse process to differentiation.

Both processes involve a philosophically tricky idea: quantities that can be made arbitrarily small. These were known as infinitesimals, and they require very careful handling. No specific number can be 'arbitrarily small', since that would make it smaller than itself. However, a number that varies can become as small as we wish. But if something varies, how can it be a number?

Suppose we know exactly where a car is located at any instant of time, and we want to work out its velocity. If, over a period of one hour, it has travelled sixty kilometres, the *average* velocity during that time is sixty kilometres per hour. But the velocity could have been faster at some times and slower at others. Reducing the time interval to one second gives a more precise estimate, the average velocity over one second. But the velocity might still have changed slightly during that time. We can approximate the instantaneous velocity, at a given moment, by finding out how far the car moves in a very short interval of time, and dividing that distance by the time it took. However small we make that interval, though, the result is only an approximation. But if you try this using a formula for the position of the car, it turns out that if you make the interval closer and closer to zero, the average velocity over that interval gets closer and closer to *some specific value*. We define that value to be the instantaneous velocity.

The usual way to do the sums requires us to divide the distance travelled by the time elapsed. Critics such as Bishop George Berkeley were quick to point out that when the elapsed time becomes zero, this fraction is $0/0$, which is meaningless. Berkeley published his criticisms in 1734 in a pamphlet *The Analyst, a Discourse Addressed to an Infidel Mathematician*, referring sarcastically to Newton's fluxions (the instantaneous velocities) as 'ghosts of departed quantities'.

Newton and Leibniz had answers to such objections. Newton employed a physical image of the interval flowing towards 0, but not actually getting there. The distance travelled flows towards 0 as well, and the average velocity flows too. What matters, Newton said, is what it flows towards. Getting there is irrelevant. So he called his method 'fluxions' – things that flow. Leibniz preferred to treat the time interval as an infinitesimal, by which he meant not a fixed nonzero quantity that can be arbitrarily small (which makes no logical sense) but a variable nonzero quantity that can *become* arbitrarily small. This is pretty much the same viewpoint as Newton's. It is, give or take some precision in terminology, the viewpoint we use today, known as 'taking a limit'. But it took several centuries to sort all this out. It's subtle. Even today, mathematics undergraduates take a while to get used to it.

Bishop Berkeley may have been unhappy about the foundations of calculus, but mathematicians are always willing to ignore philosophers, especially when the philosophers tell them to stop using a method that works perfectly well. No, the big argument about calculus was a priority dispute about who created it.

Newton had written his *Method of Fluxions and Infinite Series* in 1671, but had not published it. It finally saw the light of day in 1736 in an English translation of the Latin original by John Colson. Leibniz published on differential calculus in 1684, and on integral calculus in 1686. Newton published his *Principia* in 1687. Moreover, although many of its results depended on calculus, Newton chose to present them in a more classical geometric form, using a principle he called 'prime and ultimate ratios'. Here's how he defined equality of fluxions:

> Quantities, and the ratios of quantities, which in any finite time converge continually to equality, and before the end of that time approach nearer to each other than by any given difference, become ultimately equal.

Today's formulation of the limit concept in analysis is equivalent to this, but the meaning is made more explicit. Newton's critics never understood this definition.

Newton used geometry instead of calculus in the *Principia* to avoid getting tangled up in issues about infinitesimals, but by doing so, he missed a golden opportunity to reveal calculus to the world. Those ideas circulated informally among British mathematicians, but went largely unnoticed in the wider world. So when Leibniz became the first to publish on calculus, he caused an outcry in Britain. A Scottish mathematician named John Keill set the ball rolling by publishing an article in *Transactions of the Royal Society* accusing Leibniz of plagiarism. When Leibniz read it in 1711 he demanded a retraction, but Keill upped the ante by arguing that Leibniz had seen two letters from Newton that contained the main ideas of differential calculus. Leibniz asked the Royal Society to mediate, and it set up a committee. This came down in Newton's favour – but the report was *written* by Newton, and no one had asked Leibniz to present his own case. Big-name mathematicians in continental Europe joined in, complaining that Leibniz wasn't getting fair treatment. Leibniz stopped arguing with Keill, on the grounds that he refused to argue with an idiot. It all got out of hand.

Later historians consider the game to have been a draw. Newton and Leibniz devised their methods pretty much independently. They were to some extent aware of each other's work, but nobody stole anyone else's ideas. Various mathematicians, including Fermat and Wallis, had been circling around them for a century or more. Unfortunately, this senseless controversy caused British mathematicians to ignore what their continental cousins were doing for the next hundred years or so, which was a pity because it included most of mathematical physics.

The *Principia* built on the work of earlier scientists, especially Kepler, whose three basic laws of planetary motion led Newton to formulate his law of gravity, and Galileo, who investigated the motion of a falling body experimentally, and spotted elegant patterns in the numbers. He published his discoveries in 1590 in *On Motion*. This inspired Newton to state three general laws of motion. The first edition of the *Principia* was published in 1687; further editions, with additions and corrections, followed. In 1747 Alexis Clairaut wrote that the book 'marked the epoch of a great revolution in physics'. In the preface, Newton explained its big theme:

> Rational Mechanics will be the science of motions resulting from any forces whatsoever, and of the forces required to produce any motions ... and therefore we offer this work as mathematical principles of philosophy. For all the difficulty of philosophy seems to consist in this – from the phenomena of motions to investigate the forces of Nature, and then from these forces to demonstrate the other phenomena.

It was a bold claim, but in hindsight its optimism was fully justified. Within a century Newton's early insights had grown into a massive area of science: mathematical physics. Many of the equations developed during this period remain in use today, with applications to heat, light, sound, magnetism, electricity, gravity, vibrations, geophysics, and so on. We've gone beyond that 'classical' style of physics with relativity and quantum theory, but it's amazing how important Newtonian physics remains. And his idea of describing nature by differential equations is used throughout the sciences, from astronomy to zoology.

Book 1 of the *Principia* tackles motion in the absence of any resisting medium – no friction, no air resistance, no fluid drag. This is the simplest type of motion with the most elegant mathematics. It starts by explaining the method of first and last ratios, upon which all else rests. As explained, this is calculus in geometric disguise. It establishes early on that an inverse square law of attraction is equivalent to Kepler's laws of planetary motion. At first sight, the logical equivalence of Newton's law with Kepler's three laws suggests that all Newton achieved was to reformulate Kepler's laws in the language of forces. But there's one further feature, a prediction rather than a theorem. Newton, like Hooke before him, claims that these forces are *universal*. Every body in the universe attracts every other body. This lets him develop principles applicable to the entire solar system, and he makes a start on the problem of three bodies moving under gravitational attraction.

Book 2 tackles movement in a resisting medium, including air resistance. It develops hydrostatics – the equilibria of floating bodies – and compressible fluids. A study of waves leads to an estimate for the speed of sound in air, 1088 feet per second (331 metres per second), and how it varies with humidity. The modern figure, at sea level, is 340 metres per second. Book 2 ends by demolishing Descartes's theory of the formation of the solar system through vortices.

Book 3, subtitled *On the System of the World*, applies the principles developed in the first two books to the solar system and astronomy. The applications are astonishingly detailed: irregularities in the motion of the Moon; the movement of Jupiter's satellites, of which four were then known; comets; tides; precession of the equinoxes; and especially the heliocentric theory, which Newton formulated in a very thoughtful manner: 'The common centre of gravity of the Earth, the Sun and all the Planets is to be esteem'd the Centre of the World ... [and this centre] either is at rest, or moves uniformly forward in a right line.' By estimating the ratios of the masses of the Sun, Jupiter, and Saturn, he calculated that this common centre of gravity is very close to the centre of the Sun, with an error at most the diameter of the Sun. He was right.

✳

The inverse square law of attraction was not actually original to Newton. Kepler alluded to this type of mathematical dependence in the context of light in 1604, arguing that as a bunch of light rays spreads out, it has to illuminate a sphere whose area grows as the square of its radius. If the amount of light is conserved, brightness must be inversely proportional to the square of the distance. He also suggested a similar law for 'gravity', but what he meant was a hypothetical force exerted by the Sun that propelled planets along their orbits, and he believed it was inversely proportional to distance. Ismaël Bullialdus disagreed, arguing that this force must be inversely proportional to the square of the distance.

Gravitational attraction, its universality, and the inverse square law were all very much in the air around 1670. It is also a very natural relationship, by analogy with the geometry of light rays. In a lecture to the Royal Society in 1666 Robert Hooke said:

> I will explain a system of the world very different from any yet received. It is founded on the following positions. 1. That all the heavenly bodies have not only a gravitation of their parts to their own proper centre, but that they also mutually attract each other within their spheres of action. 2. That all bodies having a simple motion, will continue to move in a straight line, unless continually deflected from it by some extraneous force, causing them to describe a circle, an ellipse, or some other curve. 3. That this attraction is so much the greater as the bodies are nearer. As to the proportion in which those forces diminish by an increase of distance, I own I have not discovered it.

In 1679 he wrote a private letter to Newton,[3] proposing an inverse square law dependence for gravity in this sense. He was distinctly miffed when exactly that law appeared in the *Principia*, even though Newton gave him credit for it, along with Halley and Christopher Wren. We can sympathise with Hooke because, despite that, Newton got the lion's share of the credit. In part this happened because the *Principia* became so influential, but there's another reason. Newton didn't merely suggest such a law. He *deduced* it from Kepler's laws, thereby putting it on a sound scientific footing.

Hooke agreed that only Newton had given 'the Demonstration of the Curves generated thereby', that is, that closed orbits are elliptical. (The inverse square law also permits parabolic and hyperbolic orbits, but these are not closed curves and the motion does not repeat periodically.)

Nowadays we tend to see Newton as the first great rational thinker. We disregard his strong belief in God and his biblical scholarship, and we steadfastly ignore his extensive researches into alchemy, the rather mystical attempts to convert matter from one form into another. Most of his writings on alchemy were probably lost when his laboratory caught fire, and two decades of research went up in smoke. Apparently his dog was the cause: Newton is said to have scolded the animal, saying 'Oh Diamond, Diamond, thou little knowest the mischief thou hast done.'

Be that as it may, enough books and papers survive to suggest he was seeking the philosopher's stone, which would turn lead into gold. And possibly the elixir of life, the key to immortality. One title is: *Nicholas Flammel, His Exposition of the Hieroglyphicall Figures which he caused to be painted upon an Arch in St Innocents Church-yard in Paris. Together with the Secret Booke of Artephius, And the Epistle of Iohn Pontanus: Containing both the Theoricke and the Practicke of the Philosophers Stone.* Here's an excerpt:

> The spirit of this earth is ye fire in wch Pontanus digests his feculent matter, the blood of infants in wch y^e ☉ & ☽ bath themselves, the unclean green Lion w^{ch}, saith Ripley, is y^e means of joyning y^e tinctures of ☉ and ☽, the broth wch Medea poured on y^e two serpents, the Venus by meditation of w^{ch} ☉ vulgar and the ☿ of 7 eagles saith Philalethes must be decocted.

Here the symbols have the following meanings: ☉ = Sun, ☽ = Moon, ☿ = Mercury. To the modern eye, this looks like mystical nonsense. But Newton was blazing trails, and knew not where they might lead. As it happens, this one was a dead end. In notes for a lecture that he never actually gave,[4] the economist John Maynard Keynes called Newton 'the last of the magicians ... the last wonderchild to whom the Magi could do sincere and appropriate homage'. Today we mostly ignore Newton's mystical aspect, and remember him for his scientific and mathematical achievements. But by doing so, we lose sight of much that drove his remarkable mind. Before Newton, human understanding of nature was deeply entwined with the supernatural. After Newton, we came consciously to recognise that the universe runs on deep patterns, expressible through the medium of mathematics. Newton himself was a transitional figure with a foot in each world, leading humanity away from mysticism towards rationality.

8

Master of Us All
Leonhard Euler

Leonhard Euler
Born: Basel, Switzerland, 15 April 1707
Died: St Petersburg, Russia, 18 September 1783

TODAY, LEONHARD EULER PROBABLY ranks as the most important mathematician to be virtually unknown to the general public. But during his lifetime, so great was his reputation that in 1760, when Russian troops wrecked his farm in Charlottenburg during the Seven Years' War, General Ivan Saltykov promptly paid for the damage. Empress Elizabeth of Russia added another 4000 roubles, a huge amount at that time. That wasn't the end of the matter, either. Euler had been a member of the St Petersburg Academy from 1726 until, concerned about the deteriorating political state of Russia, he left for Berlin in 1741. In 1766 he returned, having negotiated a 3000 rouble per year salary, a generous pension for his wife, and promises that his sons would be looked after with lucrative positions.

Life was by no means rosy, however. Euler had been suffering from poor vision after he lost the sight of his right eye in 1738; now his left eye developed a cataract and he went almost totally blind. However, he was the fortunate possessor of an astonishing memory; he could recite the whole of Virgil's epic poem the *Aeneid*, and given a page number could tell you the first and last line on that page. Once, unable to get to sleep, Euler found the traditional method of counting sheep too trivial, and passed the time by calculating the sixth powers of all numbers up to a hundred. Several days later, he could still remember them all. His sons Johann and Christoph often acted as scribes, and so did academy members Wolfgang Krafft and Anders Lexell. Euler's grandson-in-law Nikolai Fuss helped too, becoming an official assistant in 1776. All of these people had solid mathematical backgrounds and Euler discussed his ideas with them. So successful were these arrangements that his already prodigious output increased significantly after he lost his sight.

Virtually nothing stopped Euler from working. In the 1740s, at the Berlin Academy, he undertook a huge amount of administration, supervised the botanical gardens and observatory, hired employees, handled the finances, and dealt with publications such as maps and calendars. He acted as consultant to King Frederick the Great of Prussia on the improvement of the Finlow Canal and the hydraulic system at the royal summer home of Sanssouci. The king was unimpressed. 'I wanted to have a water jet in my garden: Euler calculated the force of the wheels necessary to raise the water to a reservoir, from where it should fall back through channels, finally spurting out in Sanssouci. My mill was carried out geometrically and could not raise a mouthful of water closer than fifty paces to the reservoir. Vanity of vanities! Vanity of geometry!'[5]

Historical records show that Frederick was blaming the wrong person and the wrong subject. The king's architect for Sanssouci wrote that he wanted a lot of fountains, including a huge one spurting 30 metres into the air. The only source of water was the river Havel, 1500 metres away. Euler's plan was to dig a canal from the river to a pump, powered by a windmill. This would raise water to a reservoir that created a difference in height of about 50 metres, providing enough pressure to drive the fountain. Construction began in 1748 and proceeded without any problems until the pipeline from the pump to the reservoir was installed. This was made from strips of wood held by iron bands, like barrels. As soon as the builders started pumping water up to the reservoir, the pipes burst. Drilled tree-trunks also failed, so metal tubes had to be used, but these were too narrow to provide an adequate rate of flow. Attempts to sort it all out continued until 1756, paused during the Seven Years War, and briefly resumed. Then the King lost patience and the project was abandoned. The architect blamed Frederick, who had a habit of conceiving magnificent structures but failing to provide the necessary money. His report lists everyone responsible for the failure. Euler is not among them.

In fact, Euler's work on the design initiated the theory of hydraulic flow through pipes, analysing how the motion of the water affects the pressure in the pipe. In particular, he showed that motion causes the pressure to increase, even when there is no difference in height. Traditional hydrostatics does not predict this. Euler calculated the pressure increase, made recommendations about the pump and the pipeline, and gave explicit warnings that the builders were bunglers and the project would inevitably fail. He wrote:

> I have made calculations about the first trials at which the wooden tubes burst, as soon as the water reached a height of [20 metres]. I find that the tubes actually sustained a pressure corresponding to a [100 metre] high water column. This is a certain indication that the machine is still far from its perfection ... at all costs one has to use larger tubes.

He insisted that lead tubes should be used, not wooden ones, and that the thickness of the lead should be deduced from experiments. His advice was ignored.

Frederick never had any great respect for scientists, preferring artistic geniuses such as Voltaire. He made fun of Euler's blind eye, calling him the 'mathematical Cyclops'. When Frederick wrote about the Sanssouci fiasco, thirty years had passed, and the long-departed Euler was a convenient scapegoat. The lingering belief that he was an ivory-tower mathematician with no practical abilities is complete nonsense. He advised the government on insurance, finance, artillery, and the lottery. He was the mathematical fix-it man of his day. And all the time he kept up a steady flow of penetrating original research and textbooks that became instant classics.

He was still working the day he died. In the morning, much as usual, he gave one of his grandchildren a mathematics lesson, made some calculations about balloons on two small chalkboards, and discussed the recent discovery of the planet Uranus with Lexell and Fuss. Late that afternoon he suffered a brain haemorrhage, said 'I am dying', and expired six hours later. In his *Eulogy for Mr Euler*, Nicolas de Condorcet wrote 'Euler ceased to live and calculate.' For him, mathematics was as natural as breathing.

Euler's father Paul studied theology at Basel University and became a Protestant minister. His mother Margaret (*née* Brucker) was the daughter of a Protestant minister. But Paul also took lectures from the mathematician Jacob Bernoulli, in whose house he lived as an undergraduate, along with Jacob's brother Johann, a fellow student. The Bernoullis are the archetypal example of a mathematically talented family, and for four generations nearly all of them started out pursuing more conventional careers, but ended up doing mathematics.

Euler became a student at Basel University at the age of 13, in 1720. His father wanted him to be a pastor. By 1723 he had completed a master's degree contrasting the philosophies of Newton and Descartes, but although he was a devout Christian, theology failed to appeal, and neither did Hebrew nor Greek. Mathematics was another matter altogether: Euler loved it. And he knew how to go about making a career in it, too. His unpublished autobiographical papers include this passage:

> I soon found an opportunity to be introduced to a famous professor Johann Bernoulli ... True, he was very busy and so refused flatly to give me private lessons; but he gave me much more valuable advice to start reading more difficult mathematical books on my own and to study them as diligently as I could; if I came across some obstacle or difficulty, I was given permission to visit him freely every Sunday afternoon and he kindly explained to me everything I could not understand.

Johann quickly spotted the young man's astonishing talent, and Paul agreed to let his son change subject to study mathematics. No doubt Paul's long-standing friendship with Johann helped grease the wheels.

Euler published his first paper in 1726, and in 1727 he entered a paper for the Paris Academy's annual grand prize, which on that occasion was to find the optimal arrangement of masts on a sailing ship. Pierre Bouguer, an expert in the area, won, but Euler came second. This achievement came to the attention of St Petersburg, and when Nicolaus Bernoulli died and his position became vacant, it was offered to Euler. Aged 19, he set off for Russia, a seven-week trip: along the Rhine by boat, then by wagon, and back to a boat for the final leg.

Between 1727 and 1730 he also served as a medical lieutenant in the Russian navy, but when he was made a full professor he left the navy, and soon became a permanent member of the academy. In 1733 Daniel Bernoulli resigned from his chair at St Petersburg to return to Basel, and Euler took over as a mathematics professor. His finances had improved enough for him to marry, and he duly spliced the knot with Katarina Gsell, the daughter of an artist at the local Gymnasium (high school). Eventually the couple produced thirteen children, of which eight died in infancy, and Euler remarked that he had done some of his best work while holding a baby and surrounded by children playing.

He had persistent eyesight problems, exacerbated by a fever in 1735 which nearly killed him. As already remarked, he became nearly blind in one eye. This had very little effect on his productivity – nothing ever did. He won the Paris Academy's grand prize in 1738 and again in 1740; eventually he won it twelve times. In 1741, as Russian politics became increasingly turbulent, he left for Berlin, becoming tutor to a niece of Frederick the Great. His 25 years at Berlin produced 380 papers. He wrote books on analysis, artillery and ballistics, calculus of variations, differential calculus, the motion of the moon, planetary orbits, shipbuilding and navigation, and even popular science, in *Letters to a German Princess*.

When Pierre Louis Moreau de Maupertuis died in 1759, Euler became president of the Berlin Academy in all but the title, which he refused. Four years later King Frederick offered the presidency to Jean le Rond d'Alembert, which wasn't greatly to Euler's liking.

D'Alembert decided he didn't want to move to Berlin, but the damage was done, and Euler decided it was time to head for pastures new. Or, in this case, pastures old, for he went back to St Petersburg at the invitation of Catherine the Great. And there he ended his days, having enriched mathematics beyond measure.

It's almost impossible to convey either Euler's brilliance, or the variety and originality of his discoveries, in anything shorter than a book. Even then, it would be a challenge. But we can grasp a little of what he achieved, and gain some insight into his remarkable abilities. I'll start with pure mathematics and move on to applied, ignoring chronology to maintain some kind of flow of the ideas.

First and foremost, Euler had an amazing intuition for formulas. In his 1748 *Introduction to Analysis of the Infinite* he investigated the relation between the exponential and trigonometric functions for complex numbers, leading to the formula

$$e^{i\theta} = \cos\theta + i\sin\theta$$

From this, setting $\theta = \pi$ radians $= 180°$, it's possible to derive the famous equation

$$e^{i\pi} + 1 = 0$$

relating the two enigmatic constants e and π, and the imaginary number i. Here $e = 2 \cdot 718...$ is the base of natural logarithms and i is the symbol that Euler introduced for the square root of minus one, still standard today. Now that complex analysis is better understood, this relationship doesn't come as much of a surprise, but in Euler's day it was mind-blowing. Trigonometric functions came from the geometry of circles and the measurements of triangles; the exponential function came from the mathematics of compound interest and the calculating tool of logarithms. Why should these things be so intimately connected?

Euler's uncanny knack with formulas led to a triumph that brought him great fame at the age of 28, when he solved the Basel problem. Mathematicians had been finding interesting formulas for the sums of infinite series, perhaps the simplest being

$$1 + \frac{1}{2} + \frac{1}{4} + \frac{1}{8} + \frac{1}{16} + \frac{1}{32} + \ldots = 2$$

The Basel problem was to find the sum of the reciprocals of the squares

$$1 + \frac{1}{4} + \frac{1}{9} + \frac{1}{16} + \frac{1}{25} + \frac{1}{36} + \ldots$$

Many famous names had sought the answer without success: Leibniz, Stirling, de Moivre, and three of the most proficient Bernoullis: Jacob, Johann, and Daniel. Euler trumped them all by proving (or, at least, doing a calculation indicating – rigour was not his strong point) that the sum is *exactly* $\pi^2/6$.

A simpler infinite sum, the 'harmonic series' of reciprocals of the integers, is:

$$1 + \frac{1}{2} + \frac{1}{3} + \frac{1}{4} + \frac{1}{5} + \frac{1}{6} + \ldots$$

and this diverges – its sum is infinite. Euler, unfazed, found a highly accurate approximate formula:

$$1 + \frac{1}{2} + \frac{1}{3} + \frac{1}{4} + \frac{1}{5} + \frac{1}{6} + \ldots + \frac{1}{n} \approx \log n + \gamma$$

where γ, now called Euler's constant, is, to 16 decimal places:

$$0 \cdot 5772156649015328 \ldots$$

Euler himself calculated its value to that many decimal places. By hand.

Number theory naturally attracted Euler's attention. He took much of his inspiration from Fermat, and correspondence with his friend Goldbach, an amateur mathematician, provided further motivation. His solution of the Basel problem led him to a remarkable relation between primes and infinite series (Chapter 15). He obtained proofs of several basic theorems stated by Fermat. One was the so-called 'Little Theorem', as distinct from the Last Theorem. This states that if n is a prime and a is not a multiple of n, then $a^n - a$ is divisible by n. Innocuous as this statement may appear, it's now the starting point for some allegedly unbreakable codes, widely used on the internet. He also generalised the result to composite n, introducing the totient (or Euler) function $\varphi(n)$. This is the number of integers between 1 and n that have no prime factor in common with n. He conjectured the law of quadratic reciprocity, later proved by Gauss (Chapter 10); characterised all primes that are the sum of two squares (2, all of the form $4k + 1$, none of the form $4k + 3$), and improved Lagrange's theorem that every positive integer is the sum of four squares.

His textbooks on algebra, calculus, complex analysis, and other topics standardised mathematical notation and terminology, much of it still in use today, such as π for pi, e for the base of natural logarithms, i for the square root of minus one, the Σ notation for a sum, and $f(x)$ for a general function of x. He even brought together Newton's and Leibniz's notations in differential calculus.

I like to define a mathematician not as 'someone who does mathematics', but as 'someone who spots an opportunity for doing mathematics when no one else would'. Euler seldom missed an opportunity. Two examples kick-started the area now known as combinatorics or discrete mathematics, which is about counting and arranging finite objects.

The first, in 1735, was a puzzle about the city of Königsberg in Prussia (now Kaliningrad in Russia). Situated on the river Pregel, the city had two islands connected to each other and the banks of the river by seven bridges. The puzzle was to find a path through the city that crossed every bridge once and once only. The start and finish could be in different places. Euler proved that no such path exists, by tackling the more general question for any arrangement of islands and bridges. He proved that a path exists if and only if at most two islands are at the ends of an odd number of bridges.

Today we interpret this theorem as one of the first in graph theory, the study of networks of points connected by lines. Euler's proof was algebraic, involving a symbolic representation of the path using letters for islands and bridges. It's easy to prove his condition is necessary for a path to exist; the harder part is to prove it sufficient.

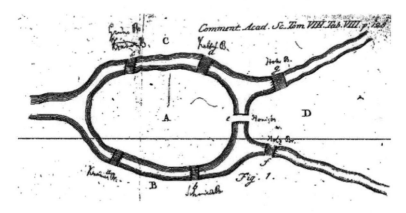

Map of the seven bridges of Königsberg, from Euler's 'Solutio problematis ad geometriam situs pertinentis'.

The second combinatorial problem, which he posed in 1782, was the 36 officers puzzle. Given six regiments, each comprising six officers of six different ranks, can they be arranged in a 6×6 square so that no row or column contains two officers in the same regiment or of the same rank? Euler conjectured that this is impossible, a result whose proof had to wait for Gaston Tarry in 1901. The underlying structure here is a Latin square, in which n copies of n symbols must be arranged in an $n \times n$ square so that each symbol occurs exactly once in each row and column. The 36 officers are required to form two 'orthogonal' Latin squares, one for the regiment and another for the rank, with all possible pairs included. Latin squares have applications to experimental design for statistical tests, and widespread generalisations known as block designs appear in several branches of mathematics. Sudoku is a variation on the theme.

＊

The results I've discussed barely scratch the surface of Euler's prodigious output of pure mathematics, but he was at least as prolific in applied mathematics and mathematical physics.

In mechanics, he systematised and advanced the state of the art for the motion of a particle in his *Mechanics* of 1736. A major innovation was the use of analysis in place of geometry, which unified the treatment of previously diverse problems. He followed this with a book about ship design, beginning with hydrostatics, which also introduced differential equations for the motion of a rigid body. This theme was developed in 1765 in *Theory of the Motion of Solid Bodies*, in which he defined a coordinate system now known as Euler angles, relating them to the body's three axes of inertia and its moments of inertia about those axes. The axes of inertia are distinguished lines representing special components of the body's spin; the corresponding moment determines the amount of spin relative to the chosen axis. In particular he solved his equations for the Euler top, a body with two equal moments of inertia.

In fluid mechanics he set up basic equations now called the Euler equations, which remain of interest even though they ignore viscosity. He studied potential theory, with applications to gravity, electricity, magnetism, and elasticity. His work on light was instrumental in the success of the wave theory, which prevailed until the appearance of quantum mechanics in 1900. Some of his results in celestial mechanics were used by the astronomer Tobias Mayer to calculate tables of the motion of the moon. In 1740 he wrote *Method for Finding Curved Lines* – the full title is much longer – initiating the calculus of variations. This seeks curves and surfaces that minimise (or maximise) some related quantity, such as length or area. All of his books are clear, elegant, and organised.

Other works cover topics such as music, map-making, and logic – there are very few areas of mathematics that *didn't* attract Euler's attention. Laplace summed up Euler's role perfectly: 'Read Euler, read Euler, he is the master of us all.'

9

The Heat Operator
Joseph Fourier

Jean Baptiste Joseph Fourier
Born: Auxerre, France, 21 March 1768
Died: Paris, France, 16 May 1830

IT WAS 1804, and mathematical physics was in the air. Johann Bernoulli had applied Newton's laws of motion, combined with Hooke's law for the force exerted by a stretched spring, to the vibrations of a violin string. His ideas led Jean le Rond d'Alembert to formulate the wave equation. This is a partial differential equation, relating the rates of change of the shape of the string relative to both space and time. It governs the behaviour of waves of all kinds – water waves, sound waves, vibrations. Similar equations had also been proposed for magnetism, electricity, and gravity. Now Joseph Fourier decided to apply the same methods to another area of physics, the flow of heat in a conducting medium. After three years of research, he produced a lengthy memoir about the propagation of heat. It was read to the Paris Institute, to mixed reactions, so a committee was set up to examine it. When the report was written, it was clear that the committee wasn't happy. They had two reasons for this, one good, one bad.

Jean-Baptiste Biot had alerted them to what he claimed was an issue in the derivation of the equation for the flow of heat. In particular, Fourier hadn't mentioned an 1804 paper of his. This was the bad reason, because Biot's paper was wrong. The good reason was that a key step in Fourier's argument, transforming a periodic function into an infinite series of sines and cosines of multiples of a given angle, had not been established with due rigour. Indeed, Euler and Bernoulli had been arguing about the same idea for years in the context of the wave equation. Fourier hastened to clarify his reasoning, but the committee remained unsatisfied.

Nevertheless, the problem was considered important and Fourier had made significant inroads into it, so the institute announced that its prize problem for 1811 would be the flow of heat in a solid. Fourier added some further results to his memoir, on cooling and radiation of heat, and submitted it. A new committee awarded him the prize, but stated the same reservation about trigonometric series:

> The manner in which the author arrives at these equations is not exempt of difficulties and his analysis to integrate them still leaves something to be desired on the score of generality and even rigour.

It was normal for the prizewinning memoir to be published, but the committee declined to do so, because of this criticism.

In 1817, Fourier was elected a member of the Paris Academy of Sciences. Five years later the secretary for the mathematics section, Jean Delambre, died. François Arago, Biot, and Fourier applied for the position, but Arago dropped out and Fourier won by a landslide. Soon after, the academy published Fourier's *Analytic Theory of Heat*, the memoir that had won the prize. This looks like Fourier having a sly dig at the committee, but it had been Delambre who sent it for publication. Still, it must have given Fourier a great deal of satisfaction.

Fourier's father was a tailor, whose first marriage produced three children. When his wife died, he remarried, and the second marriage produced no less than twelve children, of whom Joseph was the ninth. When the boy was nine years old his mother died, and his father died a year afterwards. He started his education at a school run by Auxerre cathedral's music master, studying French and Latin, at which he excelled. In 1780, aged 12, he moved on to the city's École Royale Militaire. He did well in literature, but by the age of 13 his real talent was emerging: mathematics. He read advanced texts, and within a year he had worked his way through all six volumes of Étienne Bézout's *Course of Mathematics*.

In 1787, intending to become a priest, he went to the Benedictine abbey of St Benoit-sur-Loire, but remained absorbed in mathematics. He decided not to take religious vows, left the abbey in 1789, and presented a paper on algebraic equations to the academy. A year after that he worked as a teacher at his old school. To complicate matters, he became a member of the city's revolutionary committee in 1793, writing that it was possible 'to conceive the sublime hope of establishing among us a free government exempt from kings and priests' and dedicating himself to the revolutionary cause. However, the violence of the Terror during the early days of the French Revolution repelled him, and he tried to resign. This proved politically impossible, and he was irrevocably tied up in the revolution. Factional infighting was common among the revolutionaries, all of whom had different ideas about the course that the revolution should follow, and Fourier became involved in the public support of one faction in Orléans. This led to his arrest and the prospect of Madame Guillotine. At that point Maximilien Robespierre, one of the most influential revolutionaries, was guillotined, the political atmosphere shifted, and Fourier was set free.

His mathematical career flourished under the watchful eyes of the great French mathematicians of the period. He attended the

École Normale, being among its first students when it opened in 1795. He took courses from Lagrange, who he considered to be the top scientist in Europe; Legendre, who didn't greatly impress him; and Gaspard Monge. He obtained a post at the École Centrale des Travaux Publics, later renamed the École Polytechnique. His past caught up with him, and he was arrested once more, and jailed. He was soon released, however, for reasons that remain obscure, but probably involved a flurry of behind-the-scenes activity by his students and colleagues, plus another change in the political scenario. By 1797 he had come up smelling of roses, inheriting Lagrange's chair in analysis and mechanics.

Napoleon now invaded Egypt. Fourier joined his army as a scientific adviser, alongside Monge and Étienne-Louis Malus. After Napoleon had enjoyed some early successes, Horatio Nelson destroyed the French navy in the Battle of the Nile and Napoleon was stuck in Egypt. Fourier became an administrator there, set up an educational system, and did some archaeology. He was a founder member of the mathematics division of the Cairo Institute, organising reports on the expedition's scientific discoveries. He introduced Jean-François Champollion to the Rosetta Stone, a key step in Champollion's decipherment of hieroglyphs.

In 1799 Napoleon left his army behind in Egypt and returned to Paris. Fourier followed him in 1801, and resumed his professorship. But Napoleon decided Fourier was such an able administrator that he should be made prefect of the department of Isère. It was an offer that the reluctant Fourier felt unable to refuse, so he moved to Grenoble. There he oversaw the draining of the Borgoin swamps, supervised the construction of the Grenoble–Turin highway, and worked on Napoleon's massive *Description of Egypt*, published in 1810. Fourier moved to England in 1816, but soon returned to France, becoming permanent secretary of the academy. While in Egypt, he had experienced heart problems, which continued after his return to France, with frequent bouts of breathlessness. In May 1830 he fell on the stairs, making the condition much worse, and he died shortly afterwards. His name is one of the 72 inscribed on the Eiffel Tower. But as far as mathematics is concerned, it's Fourier's time in Grenoble that had the most important consequences, because it was there that he carried out his epic research on heat.

✳

Fourier's heat equation describes, symbolically, the flow of heat in a conducting rod – say, one made of metal. If part of the rod is hotter than its surroundings, the heat spreads into nearby regions; if that part is colder than its surroundings, it gets hotter at the expense of nearby regions. The greater the temperature difference, the faster the heat spreads. The rate at which heat flows also determines how quickly the entire rod cools down. Fourier's heat equation describes how these processes interact.

Initially, different parts of the rod can be heated or cooled to different temperatures, creating a temperature profile or heat distribution. Solutions of the equation describe how this initial distribution of heat along the rod changes as time passes. The precise form of the equation led Fourier to a simple solution, in a special case. If the initial temperature distribution is a sine curve, with a maximum temperature in the middle which tails away towards the ends, then as time passes the temperature has the same profile, but this decays exponentially towards zero. What Fourier really wanted to know, however, was what happens for any initial temperature profile. Suppose, for example, that initially the rod is heated along half its length, and kept much cooler along the other half. Then the initial profile is a square wave. That's not sinusoidal.

How to get a square wave from sines and cosines. *Left*: The component sinusoidal waves. *Right*: Their sum and a square wave. The first few terms of the Fourier series are shown: additional terms make the approximation to a square wave as close as we wish.

To obtain solutions despite this obstacle, Fourier exploited an important feature of his equation: it's linear. That is, any two solutions can be added together to give another. If he could represent the initial profile as a linear combination of sine curves, then the solution would be the corresponding combination of exponentially decaying sine curves. He discovered that a square wave can be represented in this form, provided you take infinitely many sine curves and combine profiles of the form sin x, sin $2x$, sin $3x$, sin $4x$, and so on. To get an exact square wave, you need infinitely many terms like this. In fact, for a rod of length 2π, the formula is

$$\sin x + \frac{1}{3} \sin 3x + \frac{1}{5} \sin 5x + \frac{1}{7} \sin 7x + \dots$$

which is really rather pretty.

Fourier's calculations convinced him that if you use cosine terms as well, infinite trigonometric series can represent *any* initial temperature profile, however complicated, even if it has discontinuities like the one occurring in the square wave. So he could write down a solution of his heat equation in the same form. Each term decays at a different rate; the more wiggles in the sine or cosine curve, the faster its contribution decays. So the profile changes its shape as well as its size. He also derived a general formula for the terms in the series, using integration.

The committee was sufficiently impressed to award him the prize, but its members were worried about Fourier's claim that his method applies to any initial profile, even one with many jumps and other discontinuities, like the square wave only worse. Fourier appealed to physical intuition as justification, but mathematicians always worry that intuition involves hidden assumptions. Indeed, neither the method nor the problem it raised were really new. The same issue had already arisen in connection with the wave equation, causing a row between Euler and Bernoulli, and Euler had published the same integral formulas as Fourier for the series expansion, with a simpler and more elegant proof. The big difference was Fourier's assertion that his method was valid for all profiles, continuous or discontinuous, a claim that Euler had shied away from. This question was a less serious issue for waves, because a discontinuous profile would model a broken violin string, which wouldn't vibrate at all. But for heat, profiles such as the square wave had sensible physical interpretations, subject to idealised modelling assumptions.

That said, the underlying mathematical issue was the same in both cases, and at that time it remained unresolved.

With hindsight, both sides in the dispute were partly right. The basic problem is that of convergence of the series: does the infinite sum have a sensible meaning? For trigonometric series, this is a delicate issue, complicated by the need to consider more than one interpretation of 'converge'. A complete answer required three ingredients: a new theory of integration developed by Henri Lebesgue, the language and rigour of set theory as invented by Georg Cantor, and a radically new viewpoint found by Bernhard Riemann. The upshot is that Fourier's method is valid for a broad but not universal class of initial profiles. Physical intuition provides a good guide for these, and they're adequate for any sensible physical system. But mathematically, you shouldn't claim too much, for there are exceptions. So Fourier was right in spirit, but his critics had valid points too.

In the 1820s, Fourier was one of the pioneers of research into global warming. Not changes in climate caused by man-made global warming, however; he just wanted to understand why the Earth was warm enough to sustain life. To find out, he applied his understanding of heat flow to our home planet. The only obvious source of heat was the radiation that the Earth received from the Sun. The planet radiates some of this heat back into space, and the difference should account for the observed average surface temperature. But it didn't. According to his calculations, the Earth ought to be noticeably colder than it actually is. Fourier deduced that other factors must be involved, and published papers in 1824 and 1827, investigating what they might be. Eventually he decided that extra radiation from interstellar space was the most likely explanation, which turned out to be hopelessly wrong. But he also suggested (and discarded) the correct explanation: that the atmosphere can act as a kind of blanket, keeping more heat in and allowing less to radiate away.

His inspiration was an experiment carried out by the geologist and physicist Horace-Bénédict de Saussure. Investigating the possibility of using the Sun's rays to cook food, de Saussure discovered that an insulated box with three layers of glass, widely separated by layers of air, was the most efficient of his designs, and that it could reach 110°C, both on the warm plains and high in the cold mountains. Therefore the warming mechanism largely depended on the air inside the box and the effect of the glass. Fourier guessed that the Earth's atmosphere might act in the same manner as de Saussure's solar oven. The phrase 'greenhouse effect' may derive from this suggestion, but it was first used by Nils Ekholm in 1901.

Ultimately, Fourier was unconvinced that this effect was the answer he sought, in part because the box precludes convection, which transports heat over large distances in the atmosphere. He didn't appreciate the special role of carbon dioxide and other 'greenhouse gases', which absorb and emit infrared radiation in a manner that traps more heat. The precise mechanism is complicated, and the analogy with a greenhouse is misleading, because a greenhouse works by confining warm air in an enclosed space.

Fourier also developed a version of his equation for heat flow in regions of the plane, or of space, in terms of what we now call the heat operator. This combines changes in temperature at a given location with diffusion of heat into or out of its vicinity. Eventually mathematicians sorted out the sense in which Fourier series solve the heat equation, for spaces of any dimension. By then it had already become apparent that the method has far broader applications – not to heat at all, but to electronic engineering.

This is a typical example of the unity and generality of mathematics. The same technique applies to any function, not just a heat profile. It represents that function as a linear combination of simpler components, making it possible to process the data and extract information from some range of components. For example, a version of Fourier analysis is used for image compression in digital cameras – encoding an image as a combination of simple patterns based on cosine functions, which reduces the memory required to store it.

Nearly two hundred years on, Fourier's initial insight has become an indispensable tool for mathematicians, physicists, and engineers. Periodic behaviour is widespread, and whenever it happens, you can work out the corresponding Fourier series and see where it leads. A generalisation, the Fourier transform, applies to non-periodic functions. A discrete analogue, the fast Fourier transform, is one of the most widely used algorithms in applied mathematics, with applications to signal processing and high-precision arithmetic in computer algebra. Fourier series help seismologists to understand earthquakes and civil engineers to design earthquake-proof buildings. They help oceanographers to map the deep oceans and oil companies to prospect for oil. Biochemists use them to work out the structure of proteins. The Black–Scholes equation, which traders use to price stock market options, is a close relative of the heat equation. The legacy of the heat operator is almost unbounded.

10

Invisible Scaffolding
Carl Friedrich Gauss

Johann Carl Friedrich Gauss
Born: Braunschweig, Duchy of Braunschweig-Wolfenbüttel, 30 April 1777
Died: Göttingen, Kingdom of Hanover, 23 February 1855

THE YEAR IS 1796, the date 30 March. The young Carl Friedrich Gauss has been trying to decide whether to study languages or mathematics. Now he has made a very significant breakthrough, using algebraic methods to uncover a geometric construction that had gone unnoticed since the time of Euclid, more than two thousand years ago. Using only the traditional geometric instruments of ruler and compasses, he can construct a regular heptadecagon. That is, a 17-sided polygon with all sides equal and all interior angles equal. Not just approximately – that's easy – but *exactly*. Few people are given the opportunity to discover something that no one else had even suspected for two millennia; even fewer take it. Moreover, despite its esoteric nature, the mathematics is very original and of the highest beauty, though of itself it has no practical importance.

Euclid's *Elements* sets the scene. It gives constructions for an equilateral triangle, a square, a regular pentagon, a regular hexagon: regular polygons with three, four, five, and six sides. What about seven sides? No, nothing. Of course eight is easy – draw a square surrounded by a circle and cut its side in half; then extend radii through those midpoints to create four new corners on the circle. If you can construct any regular polygon, the same trick constructs one with twice as many sides. Nine? No, Euclid remains mute. Ten is easy again: just double five. Nothing about eleven. Twelve is twice six, straightforward. Thirteen, fourteen – no. Fifteen can be done by combining the constructions for three- and five-sided polygons. Sixteen: double up eight sides.

As far as Euclid goes, that's it. Three, four, five, fifteen, and all multiples of those numbers by powers of two. *Seventeen*? Crazy. Even more so, because Gauss's method makes it pretty clear that seven, nine, eleven, thirteen, and fourteen sides are impossible with ruler-and-compass constructions. But, crazy or not, it's true. There's even a simple reason (although *why* it's the reason is by no means simple). Seventeen is a prime number, and subtracting one gives sixteen, a power of two.

This combination, Gauss realises, holds the key to ruler-and-compass constructions of regular polygons. In a small notebook, he writes: 'Principia quibus innititur sectio circuli, ac divisibilitus eiusdem geometrica in septemdecim partes *etc.*' To paraphrase: the circle can be divided into seventeen [equal] parts. It's the first entry in his notebook. Later, 145 other discoveries are added, each one recorded as a brief, often cryptic, note.

Languages? Or mathematics?

No contest.

✳

Gauss was born into a poor family. His father Gerhard took a job in Brunswick (Braunschweig) as a gardener and later worked as a canal tender and a bricklayer. Gauss's mother Dorothea (*née* Benze) was so illiterate that she didn't even record her son's birth date. However, she was by no means unintelligent, and remembered that her son had entered the world on a Wednesday, eight days before the feast of the ascension. Characteristically, Gauss later used this limited information to figure out the exact day.

Gauss's intellectual brilliance quickly became apparent. When he was three years old his father was handing out wages to some labourers. Suddenly young Carl piped up to the effect: 'No, father, that's wrong, it should be –' A recalculation proved the boy right. Recognising their son's potential, Gauss's parents went to considerable lengths to help him develop it. When Gauss was eight, his schoolteacher Büttner set the class an arithmetic problem. It's often stated that this was to add the numbers from 1 to 100, but that's probably a simplification. The actual problem was probably more complicated, but along similar lines: add a lot of equally spaced numbers. The advantage of such a sum to the teacher is straightforward: there's a cunning shortcut. Avoid revealing it to your unsuspecting pupils, and you can tie them up for hours in a gigantic calculation, which they will almost certainly get wrong. The eight-year-old sat at his desk for a moment, scribbled a single number on his slate, marched up to the teacher's desk and slapped it face down. 'Ligget se,' he said, in his country dialect: 'There it lies.' This was the common way to present your answer, and implied no disrespect. As the other pupils laboured and their slates slowly piled up, Büttner watched Gauss, who waited calmly at his desk. When the slates were inspected, only Gauss's answer was correct.

Suppose the problem really was $1 + 2 + 3 + \ldots + 99 + 100$. What's the shortcut? Well, first you need the imagination to appreciate that there *is* a shortcut. Then you have to find it. The same trick also works for more complicated sums of this kind. It's widely believed that Gauss mentally grouped the numbers in pairs: one from the beginning, one from the end. Now

$$1 + 100 = 101$$

$$2 + 99 = 101$$

$$3 + 98 = 101$$

and the pattern continues (because the first number increases by one, but the second decreases to compensate) until eventually

$$50 + 51 = 101$$

There are 50 such pairs, each adds to 101, so the grand total is $50 \times 101 = 5050$.

Ligget se.

Büttner realised he had a genius on his hands, and gave Gauss the finest arithmetic text he could buy. The boy read it like a novel, and mastered it as quickly. 'He is beyond me. I can teach him nothing more,' said Büttner. But he could still help his prodigy protégé. In 1788 Gauss entered the Gymnasium, aided by Büttner and his assistant Martin Bartels. There he developed a taste for linguistics, learning High German and Latin.

Bartels knew some of the great and the good of Brunswick, and told them about Gauss's talents. The news reached the ears of Duke Karl Wilhelm Ferdinand of Brunswick-Wolfenbüttel, and in 1791, aged 14, Gauss was granted an audience. He was shy and modest – and incredibly bright. The duke, charmed and impressed in equal measure, promised to provide money for the boy's education. In 1792, sponsored by the duke, Gauss entered the Collegium Carolinum. There he developed his interest in languages, especially the classics. Gerhard declared such impractical studies a waste of time, and Dorothea put her foot down. Their son was to receive the finest education possible, and that included Greek and Latin. End of.

Gauss had been pursuing dual interests – mathematics and languages – for some time. He had independently rediscovered (without proofs) five or six important mathematical theorems, among them the law of quadratic reciprocity in number theory, which I'll describe later, and he conjectured the prime number theorem, which states that the number of primes less than x is approximately $x/\log x$ It was proved in 1896 by Charles Hadamard and Charles de la Vallée-Poussin, independently. The year 1795 saw Gauss leave Brunswick to attend the University of Göttingen. His professor Abraham Kästner mainly wrote textbooks and encyclopedias, and did no original research. Gauss was unimpressed and made his opinion crystal clear. He was heading for a career in languages when the gods of mathematics came to his rescue in spectacular fashion with the heptadecagon.

To understand how radical Gauss's discovery was, we need to go back to ancient Greece, over two thousand years ago. In the *Elements*, Euclid systematically codified the theorems of the great Greek geometers. He was a stickler for logic and demanded that everything should be proved. Well, almost everything. You had to start somewhere, with assumptions that were not proved. Euclid classified these into three types: definitions, common notions, and postulates. We now call the last two axioms.

On the basis of these assumptions, Euclid developed a large part of Greek geometry, step by step. To modern eyes, some assumptions were missing – subtle ones, such as 'if a line passes through a point inside a circle, then the line, if extended far enough, must meet the circle'. But nit-picking aside, he did a wonderful job, deducing far-reaching consequences from simple principles.

The culmination of the *Elements* was the proof that there exist precisely five regular solids; these are shapes with regular polygons as faces, arranged in the same manner at every corner. They are the tetrahedron, with four equilateral triangle faces; the cube, with six square faces; the octahedron, with eight equilateral triangle faces; the dodecahedron, with twelve regular pentagon faces; and the icosahedron, with twenty equilateral triangle faces. Now, if you're Euclid and insist on logical proofs, you can't do the three-dimensional geometry of the dodecahedron unless you've previously done the two-dimensional geometry of the regular pentagon. After all, the dodecahedron is built from twelve regular pentagons. So to get to the real meat, the regular solids, you have to deal with regular pentagons, and much else.

Among Euclid's basic assumptions is an implicit restriction on how you can construct geometric figures. Everything proceeds in terms of straight lines and circles. In effect, you can use a ruler and, as they used to say, a pair of compasses. This is a single instrument, for much the same reason that we wear a pair of trousers and cut our hair with a pair of scissors. Nowadays we often abbreviate this to 'compass', so Euclid's procedures are called ruler-and-compass constructions. His geometry is a mathematical idealisation, in which lines are infinitely thin and perfectly straight, and circles are infinitely thin and perfectly round. So Euclid's constructions are not just good enough for government work, they're *exact*: good enough for an infinitely pedantic supermind with an infinitely powerful microscope.

Gauss's approach to regular polygons is based on Descartes's discovery that geometry and algebra are two sides of the same coin, related by coordinates in the plane. A straight line is represented by an equation, which the coordinates of every point on the line must satisfy. The same goes for circles, but the equation is more complicated. If two lines or circles intersect, the points of intersection satisfy *both* equations. When you try to find these points by solving the pair of equations, everything is fairly simple for two lines. If a line meets a circle, or two circles cross, you have to solve a quadratic equation. There's a formula for this, and its key feature is taking a square root. The rest is simple arithmetic: add, subtract, multiply, divide.

Viewed through this algebraist's telescope, a ruler-and-compass construction boils down to forming a series of square roots. With a few tricks of the trade, that's the same as solving an equation whose 'degree' – the highest power of the unknown – is 2, 4, 8, 16, that is, some power of 2. Not every such equation reduces to a series of quadratics, but that power of 2 is a clue. *Which* power even tells you how many quadratics you need to string together.

Regular polygons turn into very simple equations if you use complex numbers, in which -1 has a square root. The equation for the corners of a regular pentagon, for instance, is

$$x^5 - 1 = 0$$

which is very simple and elegant. Removing the obvious real solution $x = 1$, the others satisfy

$$x^4 + x^3 + x^2 + x + 1 = 0$$

This is still quite pretty, and, crucially, of degree 4, which is a power of 2. Something similar happens for the heptadecagon, but now the equations adds up all powers of x up to the sixteenth – and 16 is again a power of 2.

On the other hand, a regular heptagon (seven sides) has a similar equation of degree 6, which is *not* a power of 2. So you definitely can't get a regular heptagon using a ruler-and-compass construction.[6] Since Euclid constructs the pentagon, its equation must reduce to a series of quadratics. With a bit of algebra, it's not hard to discover how. Thus armed, Gauss discovered that the equation for the 17-gon *also* reduces to a series of quadratics. First, $16 = 2^4$, a power of 2, which is necessary for a series of square roots to do the trick, though not always sufficient. Second, 17 is prime, which enabled Gauss to find such a series.

Any competent mathematician could follow Gauss's reasoning, once he showed the way, but no one else even suspected that Euclid had not catalogued all the possible constructible regular polygons.

Not bad for a 19-year-old.

Under the duke's sponsorship, Gauss continued to make great strides, especially in number theory. From childhood he had been a lightning calculator, able to perform complicated arithmetic rapidly in his head. In an era before computers, this ability was very useful.

It helped him make rapid advances in number theory, and his early reputation was greatly enhanced when he wrote one of the most famous research texts in mathematical history, *Disquisitiones Arithmeticae* (Arithmetical Investigations). This book did for number theory what Euclid had done for geometry two millennia earlier. It was published in 1801, thanks to a subsidy provided by the faithful duke, who was rewarded with a fulsome dedication.

One of the basic techniques in the book is a typical example of Gauss's ability to synthesise simple concepts from disorganised and complicated results. Today we call it modular arithmetic. Many key results on number theory rest on the answers to two simple questions:

When does a given number divide another?

If not, how are the two numbers related?

Fermat's distinction between $4k + 1$ and $4k + 3$ is of this type. It's about what happens when you divide a number by 4. Sometimes it goes exactly. The numbers

$$0 \ 4 \ 8 \ 12 \ 16 \ 20 \ldots$$

are exact multiples of 4. The other even numbers

$$2 \ \ 6 \ \ 10 \ \ 14 \ \ 18 \ldots$$

are not. In fact, they each leave remainder 2 when divided by 4; that is, they are a multiple of 4 plus '2 left over'. In a similar way, the odd numbers either leave remainder 1:

$$1 \ \ 5 \ \ 9 \ \ 13 \ \ 17 \ \ 21 \ldots$$

or remainder 3:

$$3 \ \ 7 \ \ 11 \ \ 15 \ \ 19 \ \ 23 \ldots$$

Before Gauss got his hands on things, the usual form of words was that these lists comprise the numbers of the forms $4k$, $4k + 1$, $4k + 2$, and $4k + 3$, putting the remainders in their usual order. Gauss said it differently: they are the lists of all numbers that are congruent to 0, 1, 2, 3 to the modulus 4. Or more briefly, thanks to Latin grammar, modulo 4.

So far that's just terminology, but what matters is structure. If you add two numbers, or multiply them, and ask which of 0, 1, 2, 3 the result is congruent to, it turns out that the answer depends only on what the original numbers are congruent to. For instance:

If you add numbers congruent to 2 and 3, the result is always congruent to 1.

If you multiply numbers congruent to 2 and 3, the result is always congruent to 2.

Let's try it out on an example. The number 14 is congruent to 2 and 23 is congruent to 3. Their sum is 37, so that should be congruent to 1. And so it is: $37 = 4 \times 9 + 1$. The product is $322 = 4 \times 80 + 2$.

This may sound a bit simple-minded, but it lets us answer questions about divisibility by 4 using just these four 'congruence classes'. Let's apply the idea to primes that are sums of two squares. Every whole number is congruent (modulo 4) to 0, 1, 2, or 3. Therefore the squares are congruent to the squares of these four numbers, that is, to 0, 1, 4, or 9. These in turn are congruent to 0, 1, 0, 1, respectively. This is a very quick and easy way to prove that every square is either of the form $4k$ or $4k + 1$, in old terminology. But there's more. Sums of two squares are therefore congruent to either $0 + 0$, $0 + 1$, or $1 + 1$; that is, 0, 1, or 2. Conspicuous by its absence is 3. Now we've proved that a sum of two squares is never congruent to 3 modulo 4. So something that looks quite tricky becomes a triviality in modular arithmetic.

If the method were limited to congruence modulo 4 it wouldn't be terribly important, but you can replace 4 by any other number. If you choose 7, for example, then every number is congruent to precisely one of 0, 1, 2, 3, 4, 5, or 6. Again you can predict the congruence class of a sum or product from those of the numbers concerned. So you can do arithmetic (hence also algebra) using congruence classes in place of numbers.

In Gauss's hands, this idea became the foundation stone of far-reaching theorems about numbers. In particular, it led him to one of his most impressive discoveries, made at the age of 18. Fermat, Euler, and Lagrange had noticed the pattern previously, but none of them had given a proof. Gauss found one, publishing it in 1796 when he was 19; altogether he found six. He privately called it the *Theorema Aureum*, the golden theorem. Its official handle, far more cumbersome and less media-friendly, is the Law of Quadratic Reciprocity. It's a tool to answer a basic question: what do perfect squares look like to a given modulus? For instance, we saw that every square (modulo 4) is either 0 or 1. These are called quadratic residues (modulo 4). The other two classes, 2 and 3, are quadratic non-residues. If instead we work (modulo 7) then the quadratic residues turn out to be

$$0 \ 1 \ 2 \ 4$$

(the squares of 0, 1, 3, 2 in that order) and the non-residues are

$$3 \ 5 \ 6$$

In general, if the modulus is an odd prime p, slightly more than half of the congruence classes are residues, and slightly less than half are non-residues. However, there's no obvious pattern regarding which number is which.

Suppose p and q are odd primes. We can ask two questions:

Is p a quadratic residue modulo q?
Is q a quadratic residue modulo p?

It's not clear these questions should bear any relation to each other, but Gauss's golden theorem states that both questions have the same answer, *unless* both p and q are of the form $4k + 3$, in which case they have opposite answers: one yes, one no. The theorem doesn't say what the answers are, just how they're related. Even so, with some extra effort, the golden theorem leads to an efficient method to decide whether a given number is or is not a quadratic residue modulo another given number. However, if it is a quadratic residue, the method doesn't tell you which square to use. Even a basic question like this one still holds deep mysteries.

The core of *Disquisitiones* is a refined theory of the arithmetic properties of quadratic forms – fancy variations on 'sum of two squares' – which has since developed into vast and complex theories, with links to many other areas of mathematics. In case this all seems terribly esoteric, quadratic residues are important in the design of good acoustics in concert halls. They tell us what shape to make the sound reflectors and absorbers on the walls. And quadratic forms lie at the heart of today's mathematics, both pure and applied.

Gauss's writings are concise, elegant, and polished. 'When one has constructed a fine building, the scaffolding should no longer be visible,' he wrote. Fair enough if you want people to admire the building, but if you're training architects and builders, close scrutiny of the scaffolding is vital. The same goes if you're training the next generation of mathematicians. Carl Jacobi complained that Gauss was 'like the fox, who erases his tracks in the sand with his tail'. Gauss wasn't alone in this practice. We saw that Archimedes needed to know the area and volume of a sphere in order to make the proofs in *On the Sphere and Cylinder* work, but he kept them up his sleeve in that book. To be fair, he did disclose the underlying intuition in *The Method*. Newton used calculus to discover many of the results in his *Principia*, and then presented them in geometric guise. Pressure on journal space, and the habits of tradition, still make much published mathematics more obscure than it need be. I'm not convinced this attitude does the profession any favours, but it's very hard to change, and there are some arguments in its favour. In particular, it's hard to follow a trail that keeps heading off the wrong way, only to retrace its steps when it gets stuck.

Gauss's academic reputation was sky-high, and he had no reason to suppose the duke would stop sponsoring him at some future date, but a permanent, salaried post would offer greater security. To obtain one, it would be a good idea to have a public reputation as well. His chance came in 1801. On the first day of that year, the astronomer Giuseppe Piazzi caused a sensation by discovering a 'new planet'. We now consider it to be a dwarf planet, but for much of the intervening time it was an asteroid. Whatever its status, its name is Ceres. Asteroids are relatively small bodies orbiting (mainly) between Mars and Jupiter. A planet had been predicted at such a distance on the basis of an empirical pattern in the sizes of planetary orbits, the Titius–Bode Law. This fitted the known planets with the exception of a gap between Mars and Jupiter, just the place for an unknown planet to lurk.

By June, a Hungarian acquaintance of Gauss's, the astronomer Baron Franz Xaver von Zach, had published observations of Ceres. However, Piazzi had managed to observe the new world only for a short distance along its orbit. When it disappeared behind the glare of the Sun, astronomers were worried they wouldn't be able to find it again. Gauss devised a new method for deriving accurate orbits from a small number of observations, and Zach published Gauss's prediction, along with several others that all disagreed with each other. In December, Zach rediscovered Ceres, almost exactly where Gauss had said it would be. The feat sealed Gauss's reputation as a mathematical maestro, and his reward was to be made director of Göttingen Observatory in 1807.

By then he was married to Johanna Ostoff, but in 1809 she died after giving birth to their second son, and then the son died too. Gauss was devastated by these family tragedies, but he kept working on his mathematics. Maybe it helped him cope by distracting him. He extended the study of Ceres into a general theory of celestial mechanics: the motion of stars, planets, and moons. In 1809 he published *Theory of the Motion of Celestial Bodies about the Sun in Conic Sections.* Less than a year after Johanna's death, he married her close friend Friederica Waldeck, usually known as Minna.

By now Gauss was enshrined as the top gun of German, indeed world, mathematics; his opinions were valued and respected, and a few words of praise or condemnation from his lips could have far-reaching effects on people's careers. On the whole, he didn't abuse his influence, and he did a lot to encourage younger mathematicians, but his outlook was very conservative. He consciously avoided anything likely to be controversial, working it out to his own satisfaction, but shying away from publication. This combination occasionally led to injustice. The most glaring example occurred in connection with non-Euclidean geometry, a story I'll postpone to the next chapter.

Gauss's works were wide-ranging. He gave the first rigorous proof of the Fundamental Theorem of Algebra, that every polynomial equation has solutions in complex numbers. He defined complex numbers rigorously as pairs of real numbers subject to specific operations. He proved a basic theorem in complex analysis, later known as Cauchy's Theorem because Augustin-Louis Cauchy obtained it independently, *and* published it. In real analysis, a function can be integrated over an interval to give the area under the corresponding curve. In complex analysis, a function can be integrated along a curved path in the complex plane. Gauss and Cauchy proved that if two paths have the same endpoints, then the value of the integral depends only on those endpoints, provided the function does not become infinite at any point inside the closed curve obtained by joining the two paths together. This simple result has profound consequences for the relation between a complex function and its singularities – the points at which it becomes infinite.

He made early steps toward topology, introducing the linking number, a topological property that can often be used to prove that two linked curves can't be unlinked by a continuous deformation. This concept was generalised to higher dimensions by Poincaré (Chapter 18). It was also the first step towards a theory of the topology of knots, a topic that Gauss also thought about, which today has applications to quantum field theory and the DNA molecule.

As director of the Göttingen Observatory, Gauss had to devote a lot of his time to the construction of a new observatory, finished in 1816. He kept busy with his mathematics too, publishing on infinite series and the hypergeometric function, an article on numerical analysis, some statistical ideas, and *Theory of the Attraction of a Homogeneous Ellipsoid*, about the gravitational attraction of a solid ellipsoid – a better approximation to the shape of a planet than a sphere. He was put in charge of a geodetic survey of Hanover in

1818, improving surveying techniques. By the 1820s, Gauss was becoming very interested in measuring the shape of the Earth. Earlier, he'd proved a result that he called his *Theorema Egregium* (remarkable theorem). This characterises the shape of a surface, independently of any surrounding space. This theorem, plus his geodetic survey, won him the 1822 Copenhagen Prize.

He now entered a difficult period of his family life. His mother was ill, and in 1817 he moved her to his own home. A position at Berlin beckoned, and his wife wanted him to accept, but he was reluctant to leave Göttingen. Then, in 1831, his wife died. The arrival of the physicist Wilhelm Weber helped him overcome his grief. Gauss had known Weber for a few years, and together they worked on the magnetic field of the Earth. Gauss wrote three major papers on the topic, developing basic results in the physics of magnetism, and applied his theory to deduce the location of the South magnetic pole. With Weber, he discovered what we now call Kirchhoff's laws for electrical circuits. They also built one of the first practical electric telegraphs, able to send messages more than a kilometre.

When Weber left Göttingen, Gauss's mathematical productivity finally began to wane. He moved into the financial sector, organising the Göttingen University widow's fund. He put the experience this gave him to good use, and made a fortune investing in company bonds. But he still supervised two doctoral students, Moritz Cantor and Richard Dedekind. The latter wrote about the calm, clear manner in which Gauss held research discussions, elaborating basic principles and then developing them in his elegant handwriting on a small blackboard. Gauss died peacefully in his sleep in 1855.

11

Bending the Rules
Nikolai Ivanovich Lobachevsky

Nikolai Ivanovich Lobachevsky
Born: Nizhny-Novgorod, Russia, 1 December 1792
Died: Kazan, Russia, 24 February 1856

FOR MORE THAN two thousand years, Euclid's *Elements* was considered the epitome of logical development. Starting from a few simple assumptions, each stated explicitly, Euclid deduced the entire machinery of geometry, one step at a time. He began with the geometry of the plane, and then proceeded to solid geometry. So compelling was Euclid's logic that his geometry was seen not just as a convenient idealised mathematical representation of the apparent structure of physical space, but as a true description of it. With the exception of spherical geometry – the geometry of the surface of a sphere, widely used in navigation as a good approximation to the form of the Earth – the default view among mathematicians and other scholars was that Euclid's geometry is the only possible geometry, so it necessarily determines the structure of physical space. Spherical geometry isn't a different *kind* of geometry; it is just the same geometry, restricted to a sphere embedded in Euclidean space. Just as plane geometry is that of a plane in Euclidean space.

All geometry is Euclidean.

One of the first to suspect that this was nonsense was Gauss, but he was reluctant to publish, believing that to do so would be to open up a can of worms. The most likely responses would be blank stares and accusations ranging from ignorance to insanity. The prudent trailblazer chooses regions of the jungle where no one will scream abuse from the treetops.

Nikolai Ivanovich Lobachevsky was more courageous, or more foolhardy, or more naive than Gauss. Probably all three. When he discovered an alternative to Euclid's geometry, just as logical as its illustrious predecessor, with its own remarkable inner beauty, he understood its importance, and put his thoughts together in *Geometriya*, finished in 1823. In 1826 he asked the Department of Physico-Mathematical Sciences at Kazan University to allow him to read a paper on the topic, and it finally saw print in an obscure journal, the *Kazan Messenger*. He also submitted it to the prestigious St Petersburg Academy of Sciences, but Mikhail Ostrogradskii, an expert applied mathematician, rejected it. In 1855 Lobachevsky, then blind, dictated a new text on non-Euclidean geometry, titled *Pangeometry*. *Geometriya* was finally published in its original form in 1909, long after his death.

His remarkable discoveries, along with those of an even more unfairly neglected mathematician, János Bolyai, are now recognised as the start of a gigantic revolution in human thought about geometry and the nature of physical space. But it is ever the fate of pioneers to be misrepresented and misunderstood. Ideas that should have been hailed for their originality are routinely denounced as nonsense, and their originators receive little recognition. Hostility is more likely; think of evolution and climate change. I sometimes feel that the human race doesn't deserve its great thinkers. When they show us the stars, prejudice and lack of imagination drag us all back into the mud.

In this case, humanity was united in the belief that geometry *must* be Euclidean. Philosophers such as Immanuel Kant went to erudite lengths to explain why that was inevitable. This belief was based on longstanding tradition, reinforced by the effort required to master the arcane arguments of Euclid, which were inflicted upon generations of schoolboys as a kind of gigantic memory test. People naturally value knowledge that comes at great effort: if Euclid's geometry were not that of real space, all that hard work would have been wasted. Another reason was the seductive line of thinking that has since been dubbed the argument from personal incredulity. *Of course* the only geometry was Euclid's. What else could it be?

Rhetorical questions sometimes get rhetorical answers, and this particular question, taken seriously, led mathematicians into very deep intellectual waters. The initial motivation was a feature of Euclid's *Elements* that looked like a flaw. Not a mistake, just something that seemed inelegant and superfluous. Euclid arranged his development of geometry logically, starting with simple assumptions that were stated explicitly and not proved. Everything else was then deduced from these assumptions, step by step. Most of the assumptions were simple and reasonable: 'all right angles are equal', for instance. But one was so complicated that it stood out like a sore thumb:

> If a line segment intersects two straight lines forming two interior angles on the same side that sum to less than two right angles, then the two lines, if extended indefinitely, meet on that side on which the angles sum to less than two right angles.

It's known as the parallel axiom (or postulate) because it's really about parallel lines. When the two straight lines are parallel, they never meet. In this case, the parallel axiom tells us that the sum of the interior angles concerned must be exactly two right angles – 180°. Conversely, if that's the case, the lines are parallel.

Parallel lines are basic and obvious: just look at ruled paper. It seems evident that such lines exist, and of course they never meet because the distance between them is the same everywhere, so it can't become zero. Surely Euclid was making a meal of something that should be obvious? A general feeling arose that it ought to be possible to prove the parallel axiom from the rest of Euclid's assumptions. Indeed, several people were convinced they'd done just that, but when independent mathematicians looked at their alleged proofs, there was always a mistake or an unnoticed assumption.

In the eleventh century Omar Khayyam made one of the earliest attempts to resolve the issue. I've mentioned his work on cubic equations (page 48), but this was by no means the only string to his mathematical bow. His *Sharh ma ashkala min musadarat kitab Uqlidis* (Explanation of the Difficulties in the Postulates in Euclid's Elements) builds on an earlier attempt by Hasan ibn al-Haytham (Alhazen) to prove the parallel axiom. Khayyam rejected this and other 'proofs' on logical grounds, and replaced them by an argument that reduced the parallel axiom to a more intuitive statement.

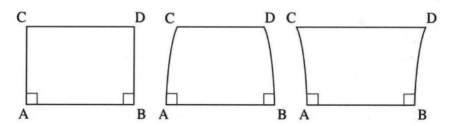

AC = BD and angles at A and B are right angles. Does DC complete the rectangle?

One of his key diagrams cuts to the heart of the problem. It can be seen as an attempt to construct a rectangle – which we might expect to be entirely straightforward. Draw a straight line, and two lines of equal length at right angles to it. Finally, join the ends of those lines to form the fourth side of the rectangle. Done!

Or is it? How do we know the result is a rectangle? In a rectangle, all four corners are right angles, and opposite sides are equal. In Khayyam's picture, we know two angles are right angles and one pair of sides is equal. What of the others?

Agreed, it *looks* as though we've drawn a rectangle, but that's because we use Euclid's geometry as a mental default. And indeed, in Euclid's geometry we can prove that CD = AB and angles C, D are also right angles. However, the deduction requires ... the parallel axiom. That's scarcely a surprise, because we expect CD to be parallel to AB. If you want to prove the parallel axiom from Euclid's other axioms, you have to prove that Khayyam has drawn a rectangle *without* appealing to the parallel axiom. In fact, as Khayyam realised, if you can find such a proof, job done. The parallel axiom itself quickly follows. Avoiding the trap of trying to prove the parallel axiom, he explicitly replaced it by a simpler assumption: 'Two convergent straight lines intersect and it is impossible for two convergent straight lines to diverge in the direction in which they converge.' And he was fully aware that this *was* an assumption.

Giovanni Saccheri took Khayyam's diagrams further, perhaps independently, but took a step backwards by trying to use them to prove the parallel axiom. His *Euclid Freed of Every Flaw* appeared in 1733. He split his proof into three possibilities, depending on whether angle C in the figure is a right angle, acute (less than a right angle), or obtuse (more than a right angle). Saccheri proved that whatever the status of angle C is in one such diagram, the same thing happens in all other diagrams of the same kind. The angles involved are either *all* right, *all* acute, or *all* obtuse. So there are three cases altogether, not three for each rectangle. That's a big step forward.

Saccheri's proof strategy was to consider the alternatives of acute and obtuse angles, aiming to refute them by deducing a contradiction. First, he assumed the angle is obtuse. This led to results that he considered incompatible with the other axioms of Euclid – case dismissed. It took him much longer to dispose of the case of the acute angle, but eventually he derived theorems that he believed contradicted the other axioms. Actually, they don't: what they contradict is Euclidean geometry, parallel axiom and all. So Saccheri thought he'd proved the parallel axiom, hence his sweeping title, whereas we now see his work as a big step towards logically consistent non-Euclidean geometries.

＊

Nikolai's father Ivan was a clerk involved in land surveys. His mother Praskovia, like his father, was a Polish emigrant. Nikolai's father died when he was seven, and his mother moved the family to Kazan in western Siberia. After finishing school there, he went to Kazan University in 1807. He started out studying medicine but soon changed to mathematics and physics. His professors included Gauss's friend and former schoolteacher Bartels.

In 1811 Lobachevsky gained a master's degree in mathematics and physics, which led to him becoming a lecturer, then extraordinary professor, then full professor by 1822. The university's administrators were backward-looking and wary of anything innovative, especially in science and philosophy. They considered such things to be some sort of dangerous spin-off from the French revolution, and a danger to the religious orthodoxy of the time. As a result, academic life was polluted, the best staff (among them Bartels) left, others were sacked, and standards declined. It wasn't the best place to be if you were about to overthrow millennia of unimaginative tradition in geometry, and Lobachevsky didn't make life any easier for himself by being outspoken and independent. Nevertheless, he kept up with his mathematical research, and his courses were models of clarity.

His administrative career, which had begun when he joined the university's buildings committee, flourished. He bought new apparatus for the physics laboratory and new books for the library. He directed the observatory, was dean of mathematics and physics from 1820 to 1825, and head librarian from 1825 to 1835. His disputes with higher authority improved when Nicholas I, who took a more relaxed attitude to politics and government, became Tsar. He removed the Curator (head) of the university, Mikhail Magnitskii, from office. His replacement Mikhail Musin-Pushkin became a staunch ally of Lobachevsky, and made him rector in 1827. The appointment, which lasted nineteen years, was a great success, with new buildings for the library, astronomy, medicine, and the sciences. He encouraged research in art and science, and student numbers increased. His quick and decisive responses ensured that a cholera epidemic in 1830 and a fire in 1842 did minimal damage, and the

Tsar sent him a message of thanks. Throughout, he delivered lectures on calculus and physics, together with general lectures for the public.

In 1832, aged 40, he married a much younger and wealthy woman, Lady Varvara Moisieva. During this period he published two works on non-Euclidean geometry: a paper on 'imaginary geometry' in 1837, and a German summary that appeared in 1840 which greatly impressed Gauss. The Lobachevskys had eighteen children, of whom seven survived to adulthood. They owned a posh house and had an extensive social life. All these left Nikolai with little money for his eventual retirement, and the marriage went poorly. His health deteriorated, and the university dismissed him in 1846, an event described as 'retirement'. His eldest son died soon after, and he started losing his sight, eventually going blind and being unable to walk. He died in 1856, in poverty, unaware that anyone would ever take any notice of his discovery of non-Euclidean geometry.

A second mathematician was equally involved in the big breakthrough: János Bolyai. His ideas saw print in 1832 as 'Appendix exhibiting the absolute science of space: independent of the truth or falsity of Euclid's axiom XI (by no means previously decided)' in his father Wolfgang's *Essay for Studious Youths on the Elements of Mathematics*. Bolyai and Lobachevsky generally get the lion's share of the credit for turning non-Euclidean geometry into a significant area of mathematics, but the prehistory of the subject includes four others who either failed to publish their ideas or were ignored when they did.

Ferdinand Schweikart investigated 'astral geometry', developing Saccheri's case of the acute angle. He sent Gauss a manuscript but never published it. He encouraged his nephew Franz Taurinus to continue the work, and in 1825 Taurinus published *Theory of Parallel Lines*. His *First Elements of Geometry* of 1826 states that the case of the obtuse angle also leads to a sensible non-Euclidean 'logarithmic-spherical' geometry. It failed to attract attention and he burnt his spare copies in disgust. One of Gauss's students, Friedrich

Wachter, also wrote about the parallel axiom, but he too was ignored.

To complicate the story further, Gauss had anticipated everyone, understanding as early as 1800 that the problem of the parallel axiom is about the internal logic of Euclidean geometry, not about real space. Ruled lines on a piece of paper can't decide the answer. Perhaps they would meet a million kilometres away if you used a big enough sheet. And perhaps, if you draw a lot of points equidistant from a straight line, the resulting line is *not straight*. Pursuing this possibility, Gauss may have started out like Saccheri, hoping to obtain a contradiction. Instead, he just obtained an increasing number of elegant, credible, mutually consistent theorems, and by 1817 he was convinced that logically consistent geometries different from Euclid's are possible. But he published nothing on the topic, commenting in a letter of 1829 that 'it may take very long before I make public my investigations on this issue: in fact, this may not happen in my lifetime for I fear the "clamour of the Boeotians".'[7]

Wolfgang Bolyai was an old friend of Gauss's, and he wrote to the great man asking him to comment (favourably, he hoped) on his son's epic research. Gauss's reply dashed his hopes:

> To praise [János's work] would be to praise myself. Indeed the whole contents of the work, the path taken by your son, the results to which he is led, coincide almost entirely with my meditations, which have occupied my mind partly for the last thirty or thirty-five years. So I remained quite stupefied. So far as my own work is concerned, of which up till now I have put little on paper, my intention was not to let it be published during my lifetime ... It is therefore a pleasant surprise for me that I am spared this trouble, and I am very glad that it is the son of my old friend, who takes the precedence of me in such a remarkable manner.

All very well, but distinctly unfair, since Gauss had published nothing. Of course, praising János's radical ideas would also risk the clamour of the Boeotians. Private praise was a bit of a cop-out, and Wolfgang and Gauss both knew it.

Lobachevsky was not aware that both Gauss and Bolyai had also tackled the problem. The parallel axiom implies the existence of a *unique* parallel to a given line through a given point, and he started by considering the possibility that this might be false. He replaced it by the existence of many such lines, where 'parallel' means 'not meeting however far extended'. He developed the consequences of this assumption in considerable detail. He didn't prove that his geometric system was logically consistent, but he failed to derive any contradictions and became convinced that none could occur. We now call his set-up hyperbolic geometry. It corresponds to Saccheri's case of the acute angle. The obtuse angle leads to elliptic geometry, very similar to spherical geometry. Bolyai studied both cases, whereas Lobachevsky limited his investigations to hyperbolic geometry.

It took a while for the validity of non-Euclidean geometry to sink in, and for its importance to be recognised. Jules Hoüel's French translation of Lobachesvky's work started the process in 1866, ten years after he died. For a while, one vital feature was conspicuous by its absence: a *proof* that denial of the parallel axiom never leads to a contradiction. This came somewhat later: there are actually three consistent geometries satisfying all of the other axioms of Euclid. These are Euclidean geometry itself; elliptic geometry, in which parallel lines do not exist; and hyperbolic geometry, where they exist but are not unique.

The consistency proof turned out to be simpler than might be expected. Non-Euclidean geometry can be realised as the natural geometry of a surface of constant curvature: positive for elliptic geometry, negative for hyperbolic geometry. Euclidean geometry is the transitional case of zero curvature. Here 'line' is interpreted as 'geodesic', the shortest path between two points. With this interpretation, all of Euclid's axioms except the parallel axiom can be proved using Euclidean geometry. If there were a logical inconsistency in either elliptic or hyperbolic geometry, it could be translated directly into a corresponding logical inconsistency in the Euclidean geometry of surfaces. Provided Euclidean geometry is consistent, so are elliptic and hyperbolic geometry.

In 1868, Eugenio Beltrami gave a concrete model for hyperbolic geometry: geodesics on a surface called the pseudosphere, which has constant negative curvature. He interpreted this result as a demonstration that hyperbolic geometry wasn't really new, just Euclid's geometry specialised to a suitable surface. In so doing, he missed the deeper logical point: the model proves hyperbolic geometry is consistent, so that the parallel axiom cannot be derived from Euclid's other axioms. Hoüel realised that in 1870 when he translated Beltrami's paper into French.

A model for elliptic geometry was easier to find. It is the geometry of great circles on a sphere, with one twist. Great circles meet at two diametrically opposite points, not at one point, so they don't obey Euclid's other axioms. To fix this up, redefine 'point' to mean 'pair of diametrically opposite points', and think of a great circle as a pair of diametrically opposite semicircles. This space, technically a sphere with diametrically opposite points identified, has constant positive curvature, inherited from the sphere.

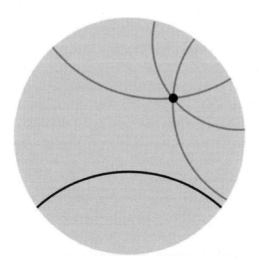

In the Poincaré disc model of hyperbolic geometry, a line (black) can have infinitely many parallels (three shown in grey) passing through a given point. The boundary of the circle is not considered to be part of the space.

Meanwhile, non-Euclidean geometry was starting to show up in other areas of mathematics, notably complex analysis, where it has connections with Möbius transformations, which map circles (and straight lines) to circles (and straight lines). Weierstrass lectured on this topic in 1870. Klein, who went along, got the message, and discussed the idea with Sophus Lie. In 1872 he wrote an influential document, the Erlangen Programme, in which he defined geometry as the study of invariants of transformation groups. This unified nearly all of the disparate geometries that were floating around by then, the main exception being Riemannian geometry for surfaces of non-constant curvature, where suitable transformation groups fail to exist. Poincaré made further advances, including his own model for hyperbolic geometry. The space is the interior of a circle, and 'straight' lines are arcs of circles meeting the boundary at right angles.

Later, hyperbolic geometry was one inspiration for Riemann's theory of curved spaces of any dimension (manifolds), which underpins Einstein's theory of gravity (Chapter 15). Its applications in modern mathematics include complex analysis, Special Relativity, combinatorial group theory, and Thurston's Geometrisation Conjecture (now theorem) in the topology of three-dimensional manifolds (Chapter 25).

12

Radicals and Revolutionaries
Évariste Galois

Évariste Galois
Born: Bourg-la-Reine, France, 25 October 1811
Died: Paris, France, 31 May 1832

ON 4 JUNE 1832 the French newspaper *Le Precursor* reported a sensational, though by no means uncommon, event:

> *Paris, 1 June.* A deplorable duel yesterday has deprived the exact sciences of a young man who gave the highest expectations, but whose celebrated precocity was lately overshadowed by his political activities. The young Évariste Galois ... was fighting with one of his

old friends ... who was known to have figured equally in a political trial. It is said that love was the cause of the combat. The pistol was the chosen weapon of the adversaries, but because of their old friendship they could not bear to look at one another and left the decision to blind fate. At point-blank range they were each armed with a pistol and fired. Only one pistol was charged. Galois was pierced through and through by a ball from his opponent; he was taken to the hospital Cochin where he died in about two hours. His age was 22. L.D., his adversary, is a bit younger.

Galois spent the night before the duel writing a summary of his mathematical researches, which centred upon the use of special sets of permutations, which he called 'groups', to determine whether an algebraic equation can be solved by a formula. He also described connections between this idea and special functions known as elliptic integrals. His results easily imply that there is no algebraic formula to solve the general quintic equation, a question that had puzzled mathematicians for centuries before Gabriel Ruffini published an almost complete but interminably lengthy proof, and Niels Henrik Abel devised a simpler one.

Several myths about Galois persist to this day, despite the efforts of historians to sort out the actual course of events. The record is patchy and sometimes contradictory. For example, who was his opponent? The newspaper article is unreliable – it gets his age wrong, for a start – and much remains obscure. The importance of his mathematics, however, is clear. The concept of a group of permutations was one of the first significant steps towards group theory. This turned out to be the key to the deep mathematics of symmetry, and is a major research area even today. Groups are now central to many areas of mathematics and indispensable in mathematical physics. They have important applications to pattern formation in many areas of physical and biological science.

Évariste's father Nicolas-Gabriel, a Republican, was mayor of Bourg-la-Reine in 1814 after Louis XVIII once more became king. His mother Adelaide-Marie (*née* Demante) was the well-educated daughter of a legal consultant. She studied religion and the classics, and she educated Évariste at home until he was 12. In 1823 he was sent to the College de Louis-le-Grand. He got first prize in Latin, became bored, and took solace in mathematics. He read advanced works: Legendre's *Elements of Geometry* and original papers by Abel and Lagrange on the solution of polynomial equations by radicals. This term refers to algebraic formulas expressing the solutions in terms of the coefficients, involving the basic operations of arithmetic and the extraction of square, cube, or higher roots. The Babylonians had solved quadratics by radicals, and the Renaissance algebraists had done the same for cubics and quartics. Now it was becoming apparent that such methods had run out of steam. Abel proved in 1824 that the general quintic – equation of the fifth degree – can't be solved by radicals, and he published an expanded proof in 1826.

Ignoring his mathematics teacher's advice, Galois took the entrance examination for the prestigious École Polytechnique a year early, without bothering to prepare for it. Unsurprisingly, he failed. In 1829 he sent a paper on the theory of equations to the Paris Academy, but it went astray. Galois interpreted this as deliberate suppression of his genius, but it could have been just sloppiness. It was a bad year all round. Galois's father killed himself during a political conflict with the village priest, who forged Nicolas's signature on malicious documents. Shortly after, Galois made a second and final attempt to get into the Polytechnique, failing again. Instead, he went to the less prestigious École Preparatoire, later renamed the École Normale. He did well in physics and mathematics, though not in literature, and graduated in both science and letters at the end of 1829. A few months later he entered a new version of his work on equations for the academy's grand prize. Fourier, the secretary, took the manuscript home, but died before making his report. Again the manuscript was lost, and again Galois saw this as a deliberate ploy to deny him the rewards his brilliance deserved. This narrative fitted neatly with his Republican views, and reinforced his determination to help foment revolution.

When opportunity knocked, Galois missed out. In 1824 Charles X succeeded Louis XVIII, but by 1830 the king was facing abdication. To avoid this, he introduced press censorship, but the people revolted in protest. After three days of chaos, a compromise candidate was agreed, and Louis-Philippe, Duke of Orléans, became king. But the director of the École Normale locked his students in. This didn't go down well with the would-be revolutionary, who wrote a blistering personal attack on the director in a letter to the *Gazette des Écoles*. Galois had signed his name, but the editor didn't print it. The director used this as an excuse to expel Galois for writing an anonymous letter. So Galois joined the Artillery of the National Guard, a militia packed with Republicans. Not long after, the king abolished it as a security threat.

In January 1831 Galois sent the academy a third manuscript on his theory of equations. After two months without any response, he wrote to the academy's president to ask what was causing the delay, but received no reply. His mental state became increasingly agitated, almost paranoid. Sophie Germain, a brilliant female mathematician, wrote about Galois to Guillaume Libri: 'They say he will go completely mad, and I fear this is true.' In April of the same year, nineteen members of the disbanded Artillery of the National Guard were tried for attempting to overthrow the government, but the jury acquitted them. At a raucous banquet of about two hundred Republicans, held to celebrate the acquittal, Galois held up a glass and a dagger. He was arrested the following day for threatening the king. He admitted his actions, but informed the court that he had proposed the toast with the words: 'To Louis-Philippe, *if he turns traitor.*' A sympathetic jury acquitted him.

In July, the academy pronounced on his submission: 'We have made every effort to understand Galois's proof. His reasoning is not sufficiently clear, sufficiently developed, for us to judge its correctness.' The referees also raised a mathematical criticism, which was entirely reasonable. They were expecting to be informed of some condition on the coefficients of the equation, which would determine whether it is soluble by radicals. Galois had proved an elegant condition, but it involved the *solutions*. Namely, each solution can be expressed as a rational function of two others. It's now clear that no simple criterion based on the coefficients exists, but no one knew that then.

Galois went ballistic. On Bastille Day, he was at the front of a Republican demonstration with his friend Ernest Duchâtelet, heavily armed and wearing his Artillery uniform. Both were illegal. The two

revolutionary comrades were arrested and incarcerated in the jail at
Sainte-Pélagie, awaiting trial. After four months, Galois was
convicted and sentenced to prison with a six-month sentence. He
passed the time by doing mathematics, and when cholera struck in
1832 he was sent to a hospital and then released on parole.

Having secured his freedom, he became obsessed with a young
woman, identified only as 'Stéphanie D' with the rest of her name
scribbled out. 'How can I console myself when in one month I have
exhausted the greatest source of happiness a man can have?' he
complained to another friend, Auguste Chevalier. He copied
fragments of the lady's letters into a notebook. One reads: 'Sir, be
assured there never would have been more. You're assuming wrongly
and your regrets have no foundation.' History has sometimes
portrayed Stéphanie as some sort of *femme fatale*, with suggestions
that it was a trumped-up 'affair of honour' that gave Galois's
enemies an excuse to challenge him to a duel. But in 1968 Carlos
Infantozzi inspected the original manuscript and reported that she
was Stéphanie-Felicie Poterin du Motel, daughter of the doctor at
Galois's lodging house. This reading is a bit controversial, but
plausible.

The police report on the duel states that it was a private dispute
about the young lady, between Galois and another revolutionary. On
the eve of the duel, Galois wrote:

> I beg patriots and my friends not to reproach me for dying
> otherwise than for my country. I die the victim of an infamous
> coquette. It is in a miserable brawl that my life is extinguished. Oh!
> why die for so trivial a thing, for something so despicable! ...
> Pardon for those who have killed me, they are of good faith.

His views of the lady would naturally be biased, but if his enemies
had engineered the whole thing, he would hardly ask for them to be
pardoned.

Who was the opponent? The record is sparse and confusing. In
his *Mémoires*, Alexandre Dumas says he was a fellow Republican,
Pescheux d'Herbinville. Which brings us back to the article in *Le
Precursor* and the enigmatic killer 'L.D.' The 'D' might refer to
d'Herbinville, but if so, the 'L' is yet another mistake in a rather
inaccurate article. Tony Rothman[8] makes a good case that 'D'

stands for Duchâtelet, though the 'L' is questionable. Many a friendship has fallen apart over a woman. The duel was with pistols – at 25 paces, according to the post-mortem report, but more like Russian roulette if *Le Precursor* is to be believed. Circumstantial evidence supports the latter, because Galois was hit in the stomach; bad luck at 25 paces but guaranteed at point-blank range. Refusing the offer of a priest, he died a day later of peritonitis and was buried in the common ditch at the cemetery of Montparnasse.

The day before the duel, Galois summarised his discoveries in a letter to Chevalier. It sketched how groups can tell us when a polynomial equation is soluble by radicals, and touched on other discoveries – elliptic functions, integration of algebraic functions, and cryptic hints whose meaning we can only guess. The letter ended:

> Ask Jacobi or Gauss publicly to give their opinion, not as to the truth, but as to the importance of these theorems. Later there will be, I hope, some people who will find it to their advantage to decipher all this mess.

Fortunately for mathematics, there were. The first person to appreciate what Galois had achieved was Joseph-Louis Liouville. In 1843, Liouville spoke to the very body that had mislaid or rejected Galois's three memoirs. 'I hope to interest the Academy,' he began, 'in announcing that among the papers of Évariste Galois I have found a solution, as precise as it is profound, of this beautiful problem: whether there exists a solution [of an equation] by radicals.' Soon, Jacobi had read Galois's papers, and – as Galois had hoped – understood their importance. By 1856, Galois theory was being taught at postgraduate level in both France and Germany. And in 1909 Jules Tannery, director of the École Normale, unveiled a memorial to Galois in his home town of Bourg-la-Reine, thanking the Mayor for 'allowing me to make an apology to the genius of Galois in the name of this school to which he entered reluctantly, where he was misunderstood, which expelled him, but for which he was, after all, one of the brightest glories.'

What, then, did Galois do for mathematics?

The short answer is that he explained the basic mathematics of symmetry, which is the concept of a group. Symmetry has become one of the central themes of mathematics and mathematical physics, underpinning our understanding of everything from animal markings to vibrating molecules, from the shape of snail shells to the quantum mechanics of fundamental particles.

The longer version is more nuanced.

His ideas were not entirely without precedent. Very few advances in mathematics are. Mathematicians mostly build on clues, hints, and suggestions from their predecessors. A convenient entry point is Cardano's *Ars Magna*, which provided solutions for algebraic equations of the third and fourth degree. Today we write these as formulas for the solutions in terms of the coefficients. The key feature of these formulas is that they build the solution using the standard operations of algebra – addition, subtraction, multiplication, division – together with square roots and cube roots. A natural guess is that the solution of a quintic (fifth-degree) equation can also be given by such a formula, most likely requiring fifth roots as well. (The fourth root is the square root of the square root, so it's superfluous.) Many mathematicians – and amateurs – sought this elusive formula. The bigger the degree, the more complicated the formulas become, so a formula for the quintic was expected to be pretty messy. But no one could find it. Gradually it began to dawn that there might be a reason for this failure: the quest was seeking a mare's nest/red herring – choose your favourite cliché – for something that doesn't exist.

This doesn't mean that *solutions* don't exist. Every quintic equation has at least one real solution, and it always has five if we allow complex numbers too, and count 'multiple' solutions correctly. But the solutions can't be encapsulated in an algebraic formula that uses nothing more esoteric than radicals.

The first serious evidence that this might be the case had emerged in the 1770s, when Lagrange wrote a huge treatise on algebraic equations. Instead of merely observing that the traditional solutions were correct, he asked why they existed at all. What features of an equation make it soluble by radicals? He unified the classical methods for degrees two, three, and four, relating them to special expressions in the solutions that behave in interesting ways when those solutions are permuted. As a trivial example, the sum of the solutions is the same, in whichever order we write them. So is the product. The classical algebraists proved that any completely symmetric expression like these can always be expressed in terms of the coefficients of the equation, without any use of radicals.

A more interesting example, for a cubic equation with solutions a_1, a_2, a_3, is the expression

$$(a_1 - a_2)(a_2 - a_3)(a_3 - a_1)$$

If we permute the solutions cyclically, so that $a_1 \to a_2$, $a_2 \to a_3$, and $a_3 \to a_1$, this expression has the same value. However, if we swap two of them, so that $a_1 \to a_2$, $a_2 \to a_1$, and $a_3 \to a_3$, the expression changes sign. That is, it is multiplied by -1 but is otherwise unchanged. Therefore its square is fully symmetric, and must be some expression in the coefficients. The expression itself is thus the *square root* of some expression in the coefficients. This helps explain why square roots enter into Cardano's formula for solving cubics. A different partially symmetric expression explains the cube roots.

Pursuing this idea, Lagrange found a unified method for solving quadratic, cubic, and quartic equations by exploiting the permutational properties of particular expressions in the solutions. He also showed that this method *fails* when you try it on the quintic. Instead of leading to a simpler equation, it leads to a more complicated one, making the problem worse. That doesn't imply that no other method can succeed, but it's a definite hint of potential trouble.

In 1799 Paolo Ruffini took the hint and published a two-volume book, *General Theory of Equations*. 'The algebraic solution of general equations of degree greater than four,' he wrote, 'is always impossible. Behold a very important theorem which I believe I am able to assert (if I do not err).' He credited Lagrange's research as inspiration. Unfortunately for Ruffini, the prospect of wading through a 500-page tome, filled with complicated algebra, just to obtain a negative result, didn't appeal to anyone else, and he was largely ignored. Leading algebraists were starting to accept that no solution was likely, which probably didn't help. Rumours that the book had errors circulated, damping down enthusiasm even more. He tried again with revised proofs, which he hoped would be easier to understand. Cauchy did write to him in 1821, saying that his book 'has always seemed to me worthy of the attention of mathematicians and which, in my judgement, proves completely the impossibility of solving algebraically equations of higher than the fourth degree.'

Cauchy's praise might have improved Ruffini's reputation, but he died within a year. After his death, a general consensus emerged that the quintic can't be solved by radicals, but the status of Ruffini's proof remained unclear. In fact, many years later, a subtle flaw was found. The gap could be patched up, making Ruffini's book even longer, but by then Abel had published a much shorter and simpler proof. Indeed, one of his results turned out to be what was needed to complete Ruffini's proof. Abel died young, of what was probably tuberculosis. The quintic seems to have been something of a poisoned chalice.

Both Ruffini and Abel picked up Lagrange's key idea: what matters is which expressions are invariant under certain permutations of the roots. Galois's great contribution was to develop a general theory, based on permutations, that applies to all polynomial equations. He didn't just prove specific equations are insoluble by radicals; he asked exactly which ones *are* soluble. His answer was that the set of permutations that preserve all algebraic relations among the roots – he called this the group of the equation – must have a particular, rather technical but precisely defined, structure. The details of this structure explain exactly which radicals will appear, when a solution by radicals exists. The absence of such a structure means there's no solution in radicals.

The structure involved is distinctly complicated, though natural from the group viewpoint. An equation is soluble by radicals if and only if its Galois group has a series of special subgroups (called 'normal') such that the final subgroup contains just one permutation, and the number of permutations in each successive subgroup is that of the previous one, divided by a prime number. The idea of the proof is that only prime radicals are needed – for instance, a sixth root is the square root of the cube root, and both 2 and 3 are prime – and each such radical reduces the size of the corresponding group by dividing by the same prime.

The Galois group of the general quartic, for instance, contains all 24 permutations of the solutions. This group has a descending chain of normal subgroups, with sizes

$$24 \ 12 \ 4 \ 2 \ 1$$

and

$$24/12 = 2 \text{ is prime}$$

$$12/4 = 3 \text{ is prime}$$

$$4/2 = 2 \text{ is prime}$$

$$2/1 = 2 \text{ is prime}$$

Therefore we can solve the quartic, and we expect to encounter square roots (from the 2s) and cube roots (from the 3s) but nothing else.

The groups for quadratic and cubic equations are smaller, and again have descending chains of normal subgroups whose sizes are related by division by a prime. What of the quintic? This has five solutions, giving 120 permutations. The only chain of normal subgroups has sizes

$$120 \ 60 \ 1$$

Since $60/1 = 60$ is not prime, there can be no solution in radicals.

Galois didn't actually write down a proof that the quintic can't be solved. Abel had already done that, and Galois knew it. Instead, he developed a general theorem characterising all equations of prime degree that *can* be solved by radicals. To show that the general quintic is not among these equations is a triviality – so trivial for Galois that he doesn't even mention it.

✳

What makes Galois significant is not so much his theorems, as his method. His group of permutations – now called the Galois group – consists of all permutations of the roots that preserve the algebraic relations between them. More generally, given some mathematical object, we can think of all transformations – perhaps permutations, perhaps something more geometric, such as rigid motions – that preserve its structure. This is called the symmetry group of that object. 'Group' here focuses on one particular aspect of Galois's groups of permutations, which he emphasised, but did not develop into a more general concept. It means that a symmetry transformation followed by another symmetry transformation always yields a symmetry transformation.

As a simple geometric example, think of a square in the plane, and transform it using rigid motions. You can slide it, spin it, even flip it over. Which motions leave the square apparently unchanged? You can't slide it; that moves its centre to a new location. You can rotate it, but only through one or more right angles. Any other angle produces a tilt that wasn't there before. Finally, you can flip it over about any of four axes: the two diagonals, and lines through the middle of opposite sides. Not forgetting the trivial transformation 'leave it alone' we get exactly 8 symmetries.

Do the same for a regular pentagon, and you get 10 symmetries; for a regular hexagon 12, and so on. A circle has infinitely many symmetries: rotation through any angle, and flipping about any diameter. Different shapes can have different numbers of symmetries. Indeed, subtler properties than the mere number of symmetries also come into play – not just how many there are, but how they combine.

Symmetry pervades every field of mathematics, from algebra to probability theory, and it has become absolutely central to mathematics and theoretical physics. Given any mathematical object, the question 'What are its symmetries?' immediately springs to mind, and the answer is often very informative. In physics, Einstein's special theory of relativity is largely about how physical quantities behave under a particular group of symmetries of physical laws, called the Lorentz group, based on the philosophical point that the laws of nature should not depend on where or when you observe it. Today, all of the fundamental particles of quantum mechanics – electrons, neutrinos, bosons, gluons, quarks – are classified and explained in terms of a single symmetry group.

Galois took a vital step along the trail that formalised symmetry as invariance under a group of transformations. It led to the abstract definition of a group, a key feature of the modern approach to algebra. Henri Poincaré once went as far as saying that groups constitute 'the whole of mathematics' stripped down to its essentials. It was an exaggeration, but an excusable one.

13

Enchantress of Number
Augusta Ada King

Augusta Ada King-Noel, Countess of Lovelace (*née* Byron)
Born: Piccadilly (now London), England, 10 December 1815
Died: Marylebone, London, 27 November 1852

IT WAS NOT A HAPPY FAMILY.

The poet Lord George Gordon Byron was convinced that he was about to become the proud father of a 'glorious boy', but he was bitterly disappointed when his wife Anne Isabella (*née* Milbanke, and known as 'Annabella') presented him with a daughter. She was named Augusta Ada – Augusta after Byron's half-sister, Augusta Leigh. Byron always called her Ada.

A month later the couple separated, and four months after that Byron left the shores of England, never to return. Lady Byron gained custody of her daughter and disdained further contact with Lord Byron, but Ada developed a more nuanced view and took an interest in his activities and whereabouts. He travelled around Europe, spending seven years in Italy, and died when Ada was eight, of a disease caught while fighting against the Ottoman Empire in the Greek War of Independence. Much later she asked to be buried next to him upon her death, a request that was duly honoured.

Annabella considered Byron to be insane, a view that was reasonable given his outrageous behaviour. Indirectly, this led to Ada's interest in mathematics. Annabella was mathematically talented and took a keen interest in the subject. Byron's abilities most definitely lay elsewhere. In an 1812 letter to his wife he wrote:

> I agree with you quite upon Mathematics too – and must be content to admire them at an incomprehensible distance – always adding them to the catalogue of my regrets – I know that two and two make four – and should be glad to prove it too if I could – though I must say if by any sort of process I could convert two and two into five it would give me much greater pleasure.

The study of mathematics was therefore, in Annabella's eyes, an ideal way to distance the child from her father. Moreover, she believed, the subject encouraged a trained and disciplined mind. To this she added music, which endowed young ladies with desirable social skills. Apparently Annabella devoted more effort to organising her daughter's education than she did to the daughter herself; mainly Ada encountered her grandmother and her nurse. In 1816 Byron wrote suggesting that it was time for Ada to 'recognise another of her relations', namely her own mother.

Ada enjoyed the advantages and disadvantages of an upper-class English upbringing, and was educated by a series of private tutors. A certain Miss Lamont interested her in geography, which she much preferred to arithmetic, so Anabella promptly insisted that one of the geography lessons should be replaced by extra arithmetic. It wasn't long before Miss Lamont was whisked away. Family members became concerned that too much pressure was being placed on the girl, with too many punishments and too few rewards. Annabella's own mathematics tutor, William Frend, was roped in to teach Ada, but he was getting old and hadn't kept up to date with the subject.

In 1829 Dr William King was brought in, but his mathematical abilities were slight. Real mathematicians know that their subject isn't a spectator sport – you have to *do* it to appreciate it. King preferred to read about it. Arabella Lawrence was engaged to tame Ada's 'argumentative disposition'. Meanwhile Ada suffered a series of health problems, including a severe bout of measles which set her back for a long time.

In 1833 Ada was presented at court, a traditional coming-of-age for members of her class. But within a few months, a far more significant event in her life occurred. She went to a party and met the original but unorthodox mathematician Charles Babbage. With this chance event, her mathematical career took a huge step forward.

The encounter was, perhaps, less fortuitous than I've indicated, because English high society moved in the same circles as prominent individuals in science, the arts, and commerce. The leading lights in these areas all knew each other, dined together in small groups, and maintained an interest in each other's activities. Ada quickly became acquainted with the luminaries of her era – the physicists Charles Wheatstone, David Brewster, and Michael Faraday, as well as the writer Charles Dickens.

Two weeks after meeting Babbage, Ada – with her mother, both as chaperone and as an interested party – visited him in his studio. The main object of attention was a fantastic, complex machine: the Difference Engine. The core of Babbage's life's work was the design and, he hoped, construction of powerful machines for performing mathematical calculations. Babbage first conceived of such a machine in 1812, when he was musing on the deficiencies of logarithm tables. Though widely used throughout the sciences and crucial to navigation, the published tables were littered with mistakes caused by human error, either when doing hand calculations or setting the results in type. The French had tried to improve accuracy by breaking the calculations down into simple steps involving only addition and subtraction, assigning each step to

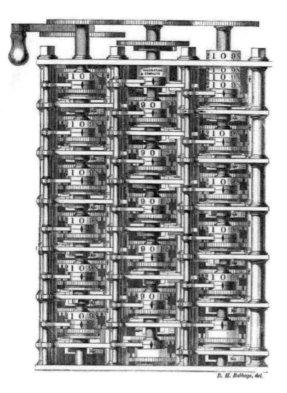

A small part of Babbage's Difference Engine.

human 'computers' trained to perform those tasks quickly and accurately, and repeatedly checking the results. Babbage realised that this approach was ideal for implementation by a machine, which, with the right design, would be cheaper, more reliable, and faster.

His first attempt in that direction, the Difference Engine, is best seen as a mechanical forerunner of the calculator; it could carry out basic operations of arithmetic. Its main role was to compute polynomial functions such as squares and cubes, or more complicated relatives, by methods based on the calculus of finite differences.

The underlying idea is simple. Patterns in these functions appear if we consider the differences between successive values. For instance, start with the cubes:

$$0 \quad 1 \quad 8 \quad 27 \quad 64 \quad 125 \quad 216$$

The differences between successive numbers go:

1   7   19   37   61   91

Take differences again:

6   12   18   24   30

and yet again:

6   6   6   6

when a simple pattern becomes obvious. (It's fairly obvious at the previous stage; less so at the one before that.) What makes this curious pattern important is the possibility of running the process backwards. Totalling series of 6s recreates the sequence immediately before that; totalling the resulting numbers gives the sequence before that one; finally, totalling that sequence yields the cubes. A similar method works for any polynomial function. You just have to be able to add. Multiplication, which looks more complicated, is superfluous.

Mechanical aids to computation were hardly a new idea. A long tradition of such aids runs throughout mathematical history, from counting on fingers to the electronic computer. But Babbage's plan was unusually ambitious. He'd gone public with the idea in a paper presented to the Royal Astronomical Society in 1822, and he extracted £1700 from the British government a year later for a pilot project. By 1842 the government's investment had risen to £17,000 – about three quarters of a million pounds (one million dollars) in today's money – with no working machine in sight. Ada and her mother had viewed a prototype, a small part of the overall plan. To make matters worse (in the government's view), Babbage then proposed a far more ambitious machine, the Analytical Engine – a genuine programmable computer, built of exquisitely engineered cogs and levers and pawls and ratchets, inspiration for the entire 'steampunk' genre of science fiction, with its mechanical versions of everything from computers to mobile phones and the internet. Unfortunately both the Difference and Analytical Engines remained pretty much that: science fiction. However, in modern times the Difference Engine was actually built, in a project headed by Doron Swade of London's Science Museum. Based on Babbage's second design, it works, and can be inspected in the museum today. Another, constructed according to Babbage's first design, is in the Computer History Museum in California. No one has yet tried to build an Analytical Engine.

✳

In 1834 Ada met one of the great female scientists, Mary Somerville, who was a close friend of Babbage. The two spent many hours discussing mathematics, and Mary lent Ada textbooks and set her problems to solve. They also talked about Babbage and the Difference Engine. The two became friends and went to scientific demonstrations, and other events such as concerts, together.

In 1835 Ada married William King-Noel, who became the First Earl of Lovelace three years later. The couple had three children, after which she returned to her first love, mathematics, under the tutelage of the noted mathematician, logician, and eccentric Augustus De Morgan, founder of the London Mathematical Society and scourge of mathematical crackpots. In 1843 she began a close collaboration with Babbage, which arose from a report on a lecture about the Analytical Engine that he'd given in Turin in 1840. Luigi Menabrea had taken notes and written them up for publication. Ada translated them from the Italian, and Babbage suggested she should add some commentary of her own. She agreed enthusiastically, and her commentary soon outstripped the original lecture.

The result was to be published in the *Taylor's Scientific Memoirs* series. At a late stage of proofreading, Babbage had second thoughts: her commentary was so good, he felt, that it would be better if she published it separately as a book. Lady King blew her aristocratic top. Most of the work already done would be wasted, the printer would be annoyed at the breach of contract – no, the idea was ridiculous. Babbage immediately backed down, as she had surely known he would. To soften the blow, Ada offered to continue writing about his work, provided no similar change of heart happened again. She also hinted that she might be able to help secure funding for the construction of the Analytical Engine, provided Babbage engaged a group of practical friends to oversee the project. Ada's mother was always complaining of ill-health; possibly what she had in mind was her likely inheritance. If so, she was disappointed, for her mother outlived her by eight years.

Ada's commentary is the main document upon which her scientific reputation rests. As well as explaining the operation of the device, it made two major contributions to what we now see as the development of the computer.

The first was to illustrate the machine's versatility. Where the Difference Engine was a calculator, the Analytical Engine was a true computer, capable of running programs[9] that could in principle calculate anything, indeed, run any specified algorithm. The idea originated with Babbage, but Ada provided a series of illustrative examples, showing how the machine could be set up to perform specific calculations. The most ambitious of these was to work out so-called Bernoulli numbers. These are named after Jacob Bernoulli, who discussed them in his *Art of Conjecturing* of 1713, one of the first books on combinatorics and probability. The Japanese mathematician Seki Kowa had discovered them earlier, but his results weren't published until after his death. They arise from the series development of the trigonometric tangent function, and occur in a variety of other mathematical contexts. They are all rational numbers (fractions), and every second Bernoulli number from the third onwards is zero; these features aside, they have no obvious pattern. The first few are:

$$1 \quad 1/2 \quad 1/6 \quad 0 \quad -1/30 \quad 0 \quad 1/42 \quad 0 \quad -1/30$$
$$0 \quad 5/66 \quad 0 \quad -691/2730$$

Despite the lack of a simple pattern, Bernoulli numbers can be calculated in turn using a simple formula. This formula was implemented in the program. I'll come back to the thorny issue of Ada's precise role here in a moment.

Her second contribution was less specific than writing programs, but much more far-reaching. Ada realised that a programmable machine can do much more than mere calculation. Her inspiration was the Jacquard loom, an extraordinarily versatile machine for weaving cloth in rich, complex patterns. The trick was to use a long chain of cards with holes punched in them, which controlled mechanical devices that activated threads of different colours, or otherwise affected the pattern of the weave. She wrote:

The distinctive characteristic of the Analytical Engine, and that which has rendered it possible to endow mechanism with such extensive faculties as bid fair to make this engine the executive right-hand of abstract algebra, is the introduction into it of the principle which Jacquard devised for regulating, by means of punched cards, the most complicated patterns in the fabrication of brocaded stuffs. It is in this that the distinction between the two engines lies. Nothing of the sort exists in the Difference Engine. We may say most aptly that the Analytical Engine weaves algebraical patterns just as the Jacquard loom weaves flowers and leaves.

This analogy then took flight. The Analytical Engine, she wrote,

might act upon other things besides number, were objects found whose mutual fundamental relations could be expressed by those of the abstract science of operations, and which should be also susceptible of adaptations to the action of the operating notation and mechanism of the engine ... Supposing, for instance, that the fundamental relations of pitched sounds in the science of harmony and of musical composition were susceptible of such expression and adaptations, the engine might compose elaborate and scientific pieces of music of any degree of complexity or extent.

Here Ada's imagination transcends that of her peers. The whole thrust of Victorian invention was a gadget for everything. One gadget to peel potatoes, another to slice boiled eggs, another to practise your riding skills without a horse ... but now, she saw that a single versatile machine could perform virtually any task. All that was needed was the right series of instructions – the program.

For this reason, Ada is often seen as the first computer programmer. She was arguably the first person to publish sample programs, although it's always possible to suggest precursors, among them Jacquard. More controversial is the extent to which the programs in her commentary are hers, rather than Babbage's. Writing in the biography *Charles Babbage, Pioneer of the Computer*, Anthony Hyman points out that three or four other people must have done similar things before: Babbage, a few assistants, and perhaps his son Herschel. Moreover, the most impressive example, the Bernoulli number program, was written by Babbage 'to save Ada the work'. Hyman concludes that 'there is not a scrap of evidence that Ada ever attempted original mathematical work'. Nevertheless, he writes that 'Ada's importance was as Babbage's interpretress. As such her achievement was remarkable.'

Against all this we must perhaps set Babbage's own words:

> We discussed together the various illustrations that might be introduced: I suggested several, but the selection was entirely her own. So also was the algebraic working out of the different problems, except, indeed, that relating to the numbers of Bernoulli, which I had offered to do to save Lady Lovelace the trouble. This she sent back to me for an amendment, having detected a grave mistake which I had made in the process.
>
> The notes of the Countess of Lovelace extend to about three times the length of the original memoir. Their author has entered fully into almost all the very difficult and abstract questions connected with the subject.
>
> These two memoirs taken together furnish, to those who are capable of understanding the reasoning, a complete demonstration – That the whole of the developments and operations of analysis are now capable of being executed by machinery.

From this scientific pinnacle, Ada's subsequent trajectory was largely downwards. Something of a wild child, she was strong-willed and impulsive. A series of affairs with gentleman friends was hushed up, and her husband had a hundred or more compromising letters of hers destroyed. A liking for wine went out of control and she also indulged in opium. She became an inveterate gambler and left debts of £2,000 on her death. The gambling may even have stemmed from a misguided attempt to raise money for the Analytical Engine.

Her health, never good, declined, and she died of cancer at the age of 37. To the end, her mind remained active and her intelligence acute. She intuitively grasped the big picture, yet she had complete mastery of the details. In 1843 Babbage summed her up: 'Forget this world and all its troubles and if possible its multitudinous Charlatans – every thing in short but the Enchantress of Number.' Nothing ever made him change that opinion.

14

The Laws of Thought
George Boole

George Boole
Born: Lincoln, England, 2 November 1815
Died: Cork, Ireland, 8 December 1864

WHEN GEORGE BOOLE WAS 16 he decided to become an Anglican clergyman, but his father's shoemaking business failed, flinging him headlong into the role of family provider. A career in the Church was no longer sensible, because the English clergy were poorly paid. He was also becoming increasingly unsure about the doctrine of the Holy Trinity, veering strongly towards the more literally monotheistic views of the Unitarians, a sect whose stance has been characterised as 'belief in at most one God'. This made it impossible for him to sign up to the Thirty-Nine Articles of the Church of England without going against his conscience.

The most – perhaps only – suitable position, given his background and talents, was in teaching, and in 1831 he took up the post of usher (assistant teacher) at Mr Heigham's School in Doncaster, some forty miles from his home town of Lincoln. In the middle of the nineteenth century this was quite a distance, and he was homesick; one letter says wistfully that nobody in Doncaster could make gooseberry pies as good as his mother's. It might have been no more than an attempt to pay her a compliment, but Boole complained about his lot for much of his career. His Unitarian tendencies, combined with a habit of solving mathematical problems in chapel on Sundays, outraged some of his students' parents, who were staunch Methodists. They complained to the headmaster and their sons prayed for Boole's soul at prayer meetings. Heigham, though happy with Boole's performance as a teacher, reluctantly fired him, replacing him with a Wesleyan.

Gooseberry pies and sectarian squabbles notwithstanding, Boole started delving even deeper into mathematics, pursuing his studies without the advice of a tutor. At first he relied on a public service, the circulating library, which had numerous textbooks at surprisingly advanced levels, but the library was disbanded, so Boole was forced to buy his own texts. Mathematical textbooks, as it happened, provided the maximum stimulation for the minimum outlay, and he purchased Sylvestre Lacroix's *Differential and Integral Calculus*. A fellow teacher wrote that during an hour set aside for the teaching of writing, from which Boole was excused: 'Mr Boole is profoundly happy; for an hour at least he can study old Lacroix without interruption.'

Later, Boole became convinced that he'd made a mistake buying a text as outmoded as Lacroix's, but studying it gave him confidence in his own abilities. One consequence was an idea that struck him, briefly but forcibly, in early 1833, while out walking across a farmer's field. Namely, the possibility of expressing logic in symbolic form. He didn't develop the idea until many years later, publishing his first book on the topic in 1847: *The Mathematical Analysis of Logic, Being an Essay Towards a Calculus of Deductive Reasoning*. Augustus de Morgan, with whom Boole had frequent correspondence, encouraged him to prepare a more extensive, better thought out book. His interests and Boole's overlapped substantially. Boole took the advice, and in 1854 his masterwork duly appeared: *An Investigation into the Laws of Thought, on Which are Founded the Mathematical Theories of Logic and Probabilities*. In this work he created mathematical logic, setting up what eventually became a theoretical basis for computer science.

Boole's father John came from a long-established Lincolnshire family of farmers and traders, 'the best thatchers and the most reading men' in the tiny village of Broxholme. He became a shoemaker, and left for London, hoping to make his fortune. Working alone in a dark cellar, he staved off depression by studying French, science, and mathematics, especially the design of optical instruments. He met and married Mary Joyce, a ladies' maid, and after six months they moved to Lincoln, where John opened a cobbler's shop. They wanted a child, but it was ten years before one arrived; they named him George. A girl and two boys followed in short order.

John was much more interested in making telescopes than shoes, so the business stumbled along, but the Booles made a living by renting rooms to lodgers. George grew up in a scientific atmosphere and had an enquiring mind. His father taught him English and mathematics. George loved mathematics and finished a six-volume geometry text by the time he was eleven (his father wrote this in pencil inside the book). He read widely and had an almost eidetic memory, able to recall any required fact instantly.

At the age of 16, Boole became a teacher at Heigham's School. After two more teaching positions he set up his own school in Lincoln at the age of 19; then he took over Hall's Academy in Waddington. His family joined him to help run the school. Boole never lost sight of higher mathematics, and was reading Laplace and Lagrange. He opened a boarding school in Lincoln, and began publishing research in the newly founded *Cambridge Mathematical Journal*.

In 1842 he began a lifelong correspondence with fellow spirit De Morgan. In 1844 he won the Royal Society's Royal Medal, and in

1849, buoyed by his growing reputation, he was appointed as the first professor of mathematics at Queen's College, Cork, Ireland. There he met his future wife Mary Everest (niece of George Everest, who completed the first major survey of India, leading to the world's highest mountain being named after him) in 1850. They married in 1855, and had five daughters, all remarkable: Mary, who married the mathematician and author Charles Howard Hinton, a brilliant rascal; Margaret, who married the artist Edward Ingram Taylor; Alicia, who was influenced by Hinton and did significant research on four-dimensional regular solids; Lucy, the first female chemistry professor in England; and Ethel, who married the Polish scientist and revolutionary Wilfrid Voynich and wrote the novel *The Gadfly.*

Among Boole's early work is a simple discovery that led to invariant theory, an area of algebra that became a very hot topic indeed. In the study of algebraic equations, a formula can sometimes be simplified if its variables are replaced by suitable expressions in a new set of variables. Solve this simpler equation to find the values of the new variables, then work backwards to deduce the values of the original ones. This is how the Babylonian and Renaissance solutions of equations worked.

An especially important class of changes of variables occurs when the new variables are linear combinations – expressions like $2x - 3y$, involving no higher powers or products of the old variables x and y. A general quadratic form

$$ax^2 + bxy + cy^2$$

in two variables can be simplified in this manner. An important quantity in the theory of such forms is the 'discriminant' $b^2 - 4ac$. Boole discovered that after a linear change of variables, the discriminant of the new quadratic form is that of the original, multiplied by a factor that depends only on the change of variables.

This apparent coincidence has a geometric explanation. It really *is* a coincidence, in the sense that two features that usually are separate coincide. If we equate the quadratic form to zero, its solutions define two (possibly complex) lines ... *unless* the discriminant is zero, in which case we get the same line *twice*. The quadratic is then the square $(px + qy)^2$ of a linear form. A coordinate change is a geometric distortion, and it carries the original lines to the corresponding ones for the new variables. If the two lines coincide for the original variables, they therefore coincide for the new ones. So the discriminants must be related in such a manner that if one vanishes, so does the other. Invariance is the formal expression of this relationship.

Boole's observation about the discriminant seemed little more than a curiosity, until a few mathematicians, the most prominent being Arthur Cayley and James Joseph Sylvester, generalised his insight to forms of higher degree in two or more unknowns. These expressions also possess invariants, which also determine significant geometric features of the associated hypersurface, defined by equating the form to zero. An entire industry emerged, in which mathematicians won their spurs by calculating invariants of ever more complicated expressions. Eventually Hilbert (Chapter 19) proved two fundamental theorems that pretty much killed the topic off until it was revived in a more general form. It remains of interest today, with important applications to physics, and had been given a new lease of life by the development of computer algebra.

The research that made Boole a household name among mathematicians and computer scientists – and in any household whose searches on Google enter into the heady realm of Boolean searches – had increasingly been occupying his thoughts. Boole always sought the inner simplicities that underpin mathematical concepts. He liked to formulate general principles, cast them in symbolic form, and let the symbols do the thinking. *The Laws of Thought* carried out this programme for the rules of logic. Its big idea was to interpret these rules as algebraic operations on symbols representing statements. Because logic is not the same as arithmetic, some of the usual algebraic laws might not apply; on the other hand, there might be new laws, which don't apply to arithmetic.

The upshot, known as Boolean algebra, makes it possible to prove logical statements by performing algebraic calculations.

The book opens with a rather deferential preface, and locates the discussion in the context of existing philosophy. Then Boole moves on to the real meat, the mathematics, with a discussion of the use of symbols. He specialises to symbols (he calls them 'signs') that represent logical statements, focusing in particular on the general laws they obey. He tells us that he will represent a class, or collection, of individuals, to which a particular name applies, by a single letter, such as x. If the name is 'sheep', then x is the class of all sheep. A class may be described by an adjective, such as 'white', in which case we might have a class y of all white things. The product xy then denotes the class of all things having both properties, that is, all white sheep. Since this class does not depend on the order in which the properties are stated, $xy = yx$. Similarly, if z is a third class (Boole's example is x = rivers, y = estuaries, z = navigable) then $(xy)z = x(yz)$. These are the commutative and associative laws of standard algebra, interpreted in this new context.

He notes one law, vital to the whole enterprise, that is not true in ordinary algebra. The class xx is the class of all things that have the property defining x and the property defining x, so it must be the same as x. Therefore $xx = x$. For example, the class of things that are sheep *and* are sheep is simply the class of all sheep. This law can also be written as $x^2 = x$, and it represents the first point at which the laws of thought depart from those of ordinary algebra.

Next, Boole moves on to signs 'whereby we collect parts into a whole, or separate a whole into its parts'. For example, suppose that x is the class of all men and y is the class of all women. Then the class of all adults – either men *or* women – is denoted $x + y$. Again there's a commutative law, which Boole makes explicit, and an associative law, which comes under the umbrella statement that the 'laws are identical' with those of algebra. Since, for example, the class of European men or women is the same as that of European men or European women, the distributive law $z(x + y) = zx + zy$ also holds, with z being the class of all Europeans.

Subtraction can be used to remove members from a class. If x

represents men and y Asiatics, then $x - y$ represents all men who are not Asiatics, and $z(x - y) = zx - zy$.

Perhaps the most striking feature of this formulation is that it's not overtly about logic. It's about set theory. Instead of manipulating logical *statements*, Boole works with the corresponding *classes*, comprising those things for which the statement is true. Mathematicians have long recognised a duality between these concepts: each class corresponds to the statement 'belongs to the class'; each statement corresponds to 'the class of things for which the statement is true'. This correspondence translates properties of classes into properties of the associated statements, and conversely.

Boole introduces this idea by way of a third class of symbols 'by which relation is expressed, and by which we form propositions'. For example, represent stars by x, suns by y, and planets by z. Then the statement 'the stars are the suns and the planets' can be stated as $x = y + z$. So propositions are *equalities* between expressions involving classes. It's an easy deduction that 'the stars, except the planets, are suns'; that is, $x - z = y$. 'This,' Boole tells us, 'is in accordance with the algebraic rule of transposition.' Al-Khwarizmi would have recognised this rule as *al-muqabala* (page 31).

The upshot of all this is that the algebra of classes obeys the same laws as ordinary algebra with numbers, plus the strange new law $x^2 = x$. At this point Boole has a very clever idea. The only *numbers* obeying that law are $0 = 0^2$ and $1 = 1^2$. He writes:

> Let us conceive, then, of an Algebra in which the symbols x, y, z, etc. admit indifferently of the values 0 and 1, and of these values alone. The laws, the axioms, and the processes, of such an Algebra will be identical in their whole extent with the laws, the axioms, and the processes of an Algebra of Logic. Difference of interpretation will alone divide them.

This enigmatic statement can be interpreted as referring to functions $f(x, y, z, \ldots)$, defined on some list of symbols, that take only the values 0 (false) or 1 (true) We now call these Boolean functions. One delightful theorem deserves mention. If $f(x)$ is a function of one logical symbol, Boole proves that

$$f(x) = f(1)x + f(0)(1 - x)$$

A more general equation of the same kind is valid for any number of symbols, leading to systematic methods for manipulating logical propositions.

Armed with this principle and other general results, Boole develops numerous examples, and shows how his reasoning applies to topics that would interest the readers of his time. These include Samuel Clarke's *Demonstration of the Being and Attributes of God*, which consists of a series of theorems, proved using observational facts and various 'hypothetical principles, the authority and universality of which are supposed to be recognised *a priori*' and the *Ethics* of Benedict Spinoza. Boole's aim here is to explain exactly which assumptions are involved in the deductions made by these authors. His quasi-Unitarian beliefs may also be making a cameo appearance.

Previous analysis of logic had been verbal, with a few symbolic mnemonics. Aristotle discussed syllogisms – arguments along the lines

> All men are mortal
> Socrates is a man
> Therefore Socrates is mortal

with variations on the use of 'all' and 'some'. Medieval scholars classified syllogisms into 24 types, giving them mnemonic names. For example, *Bocardo* refers to syllogisms of the form

> Some pigs have curly tails
> All pigs are mammals
> Therefore some mammals have curly tails

Here the vowels in 'bOcArdO' indicate the format, where O = 'some' and A = 'all'. The same convention was used to name other types of syllogism. But no systematic symbolic *notation* for logic was introduced before Boole. Notice that if we replace 'some' by 'all', obtaining

All pigs have curly tails
All pigs are mammals
Therefore all mammals have curly tails

the new syllogism becomes illogical. On the other hand,

All pigs are mammals
All mammals have curly tails
Therefore all pigs have curly tails

is a logically correct deduction – even though in reality the second statement is false. As it happens, the concluding statement is, give or take the odd special breed of pig, true.

To explain how his symbolism relates to classical logic, Boole reinterprets Aristotle, showing how the validity (or not) of each type of syllogism can be proved symbolically. For instance, let

p = the class of all pigs
m = the class of all mammals
c = the class of all creatures with curly tails

Then the final syllogism above translates into Boole's symbolism as $p = pm$ and $m = mc$, therefore $p = pm = p(mc) = (pm)c = pc$.

The rest of the book develops analogous methods for calculating probabilities, and ends with a general discussion of 'the nature of science and the constitution of the intellect'.

Boole wasn't particularly happy at Cork. In 1850, after returning from a delightful holiday in Yorkshire, he asked De Morgan: 'If you should hear of any situation in England that would be likely to suit me, to let me know of it,' and remarked that 'I no longer feel as if I could make this place my home.' One source of discontent was the authoritarian and religiously orthodox university administration, which cracked down on anyone who disagreed with it. The professor of modern languages, Raymond de Vericour, had just been suspended because of anti-Catholic remarks in a book he had written. The university's Council, under President Robert Kane, acted in such haste that it contravened the university's own statutes, and de Vericour's appeal secured his position again. Boole sided with de Vericour, but kept his head down. In 1856, further high-handed action by Kane, aimed at Boole's wife's uncle John Ryall, led Boole to write a stinging letter to the *Cork Daily Reporter*. Kane's reply was long and rather defensive, and Boole responded with yet another letter. Finally the government opened an official enquiry, denounced Kane for not spending enough time at the college, and censured both him and Boole for airing their dispute in public. Kane moved his family to Cork and everything settled down, though from then on the two maintained a cold politeness towards each other.

In 1854 Boole's mind turned to positions that had become available in Melbourne, Australia, but late in 1855 he dropped the idea completely when Mary Everest accepted his proposal of marriage. The Booles rented a large house overlooking the sea, close to the newly opened railway line, making it easy for George to commute – though he did at one point ask the college to set its clocks back by 15 minutes to make it easier for him and the students to travel by a later train. The college rejected the proposal. His eccentricity showed in other ways: he once arrived at a lecture thinking about a problem, paced up and down for an hour contemplating it, while the ranks of students on their benches felt unable to interrupt, and then left, complaining to his wife that a 'most extraordinary thing happened today. None of my students came to my lecture.'

Late in 1864 he walked from his house to the college, a distance of about 4–5 kilometres, in a heavy downpour. He fell ill with a severe cold, which affected his lungs. Mary Boole, a devotee of homeopathy, engaged a homeopath to treat him. It didn't work, and he died of pleuro-pneumonia. Ethel Voynitch, his fifth daughter, wrote:

> In Aunt Mary's [Boole's sister] view at least, the cause of Father's early death was ... the Missus's [Mary Boole's] belief in a certain crank doctor who advocated cold water cures for everything... The Everests do seem to have been a family of cranks and followers of cranks.

Ironically, Boole himself considered homeopathy ineffective. In 1860

De Morgan told him that he believed that homeopathy had cured his pleurisy, but Boole was sceptical:

> I have witnessed pleurisy and its former mode of treatment... One would say beforehand that homeopathy could have no effect on such a disease... The moral is – if you are ever attacked with an inflammation and homeopathy does not [work] ... do not sacrifice your life to an opinion ... but call in some accredited [doctor].

The area of mathematical logic opened up by Boolean algebra is now known as the propositional calculus. It goes back to the fifth century BC, when Euclid of Megara (not to be confused with the geometer Euclid of Alexandria) initiated what later became the Stoic school of logic. A key feature of Stoic logic is the use of conditional reasoning, of the form 'if A then B'. Diodorus and Philo of Megara disagreed on a fundamental issue, which continues to flummox mathematics students today. Namely, given the truth or falsity of A and B, when is the implication 'if A then B' true? Note that what's under discussion is not the truth of either A or B, but that of the *deduction* of A from B. Philo's answer was that the implication is false if A is true and B is false, but otherwise it's true. In particular, it is true whenever A is false. Diodorus's answer was different: whenever A cannot lead to a false conclusion. That boils down to 'both A and B are true'.

Today's mathematical logicians side with Philo. The counterintuitive case is when A is false. If B is also false, it seems reasonable to accept the inference 'if A then B' as valid. In particular, 'If A, then A' seems a reasonable inference, whatever the truth value of A may be. If B is true, or its current status is unknown, however, it may seem unreasonable to accept its deduction from a falsehood. For example, the statement

If $2 + 2 = 5$ then Fermat's Last Theorem is true

is considered *true* – whether Fermat's Last Theorem is true or false. (That doesn't lead to an easy proof of Fermat's Last Theorem, because to deduce that you must first prove that $2 + 2 = 5$, which is impossible if mathematics is consistent. This is why Philo's convention does no harm.) To illustrate the reasoning behind this convention, consider the following two deductions:

If $1 = -1$ then $2 = 0$ [add 1 to each side]
If $1 = -1$ then $1 = 1$ [square each side]

Both deductions are logically sound, by the reasoning in brackets. The first takes the form

If (false statement) then (false statement)

and the second takes the form

If (false statement) then (true statement)

So valid reasoning, starting from a false statement, can lead either to a false one or a true one.

Another approach that gives the same result is to ask what's needed to *disprove* an implication 'if A then B'. That is, prove it false. For example, to disprove

If pigs had wings, they'd fly

we must exhibit a winged pig that can't fly. So 'if A then B' is false if A is true and B is false, but in all other cases, the implication is true, since we can't prove it false.

This argument isn't a proof. It's motivation for the convention used in predicate logic. In modal logic, conditionals are handled differently. For example, the statement about winged pigs would be considered true, subject to the wings being functional for flight. But the similar statement

If pigs had wings, they'd play poker

would be considered false, since – even hypothetically – possessing wings does not enhance one's poker-playing abilities. In contrast, the latter statement is considered to be true in predicate logic, because pigs don't have wings. Poker doesn't come into it. This illustrates some of the difficulties that Boole and other early logicians were grappling with, and it warns us not to assume that today's conventions are necessarily the last word.

The use of Boolean algebra, or propositional calculus, in

computing stems from representing numerical and other data using the binary system, which requires only the digits 0 and 1. In their simplest manifestations, these correspond to 'no electrical voltage' and 'some electrical voltage' (at a specified level, for example 5 volts). In today's computers, all data, programs included, are encoded in binary. The data are manipulated by electronic circuits that, among other things, implement the operations of the propositional calculus – essentially Boolean algebra. Each such operation corresponds to a 'gate', and as an electrical signal or signals pass through the gate, the output depends on the input(s) according to the logical operation concerned.

This idea was pioneered by the guru of information theory, Claude Shannon. Operations on digital data performed by computers can be implemented by suitable electronic circuits, made from logic gates. So Boolean algebra is the natural mathematical language for this aspect of computer design. Early electronic engineers implemented these operations in relay circuits, and then valve (vacuum tube) circuits. With the invention of the transistor, valves were replaced by solid-state circuitry; today we use complex arrays of incredibly tiny circuits deposited on silicon chips.

Boole's formalisation of logic in symbolic terms opened up a new world, paving the way for the digital era whose fruits we now enjoy. And, frequently, curse, for we've not yet fully mastered our new technology, despite handing it ever greater control of everything in our lives.

15

Musician of the Primes
Bernhard Riemann

Georg Friedrich Bernhard Riemann
Born: Breselenz, Kingdom of Hanover, 17 September 1826
Died: Selasca, Italy, 20 July 1866

BERNHARD RIEMANN HAD SHOWN immense mathematical talent, technical mastery, and originality from the age of 20. Moritz Stern, one of his tutors, later said that 'he already sang like a canary'. His other tutor, Gauss, seemed less impressed, but the courses Gauss was teaching were elementary, unlikely to showcase the student's true abilities. Soon, even Gauss understood that Riemann was unusually able, and supervised his doctoral thesis. The topic was dear to Gauss's heart: complex analysis. Gauss commented on the 'gloriously fertile originality' of the work and arranged an entry-level position for Riemann at Göttingen University.

In Germany, the next step after the PhD was Habilitation, a more advanced degree requiring deeper research that opened up a proper academic career by entitling its holder to become a

Privatdozent, able to give lectures and charge fees. Riemann had spent two and a half years making big strides in the theory of Fourier series (Chapter 9). The research had gone well, but now he was beginning to think he'd bitten off more than he could chew.

The problem wasn't the work on Fourier series. That was done and dusted, and Riemann was confident of its quality and accuracy. No, the problem was the final step in qualifying for Habilitation. The candidate had to deliver a public lecture. He had proposed three topics: two on the mathematical physics of electricity, a subject he had also studied under Wilhelm Weber, and, more daringly, one on the foundations of geometry, where he had some interesting but rather half-baked ideas. The choice from these three topics was up to Gauss, who at that time was working with Weber and deeply interested in electricity. What Riemann didn't take into account was that Gauss was also deeply interested in geometry, and wanted to hear what Riemann had to say about it.

So now Riemann was working his socks off trying to develop his vague ideas about geometry into something that would make a real impression on the greatest mathematician of the age, in an area that luminary had been thinking about for much of his life. His starting point was a result of which Gauss was especially proud, his *Theorema Egregium* (page 109). This specifies the shape of a surface without reference to any surrounding space, and it inaugurated the subject of differential geometry. It led Gauss to study geodesics – shortest paths between points – and curvature, which quantifies how much the surface bends compared to the ordinary Euclidean plane.

Riemann planned to generalise Gauss's entire theory in a radical direction: spaces of any dimension. Mathematicians and physicists were just beginning to appreciate the power and clarity of geometric thinking in 'spaces' with more than the usual two or three dimensions. Underlying this counterfactual viewpoint was something entirely sensible, the mathematics of equations in many variables. The variables play the role of coordinates, so the more variables there are, the greater is the dimension of this conceptual space.

Riemann's efforts to develop this notion led him to the brink of a nervous breakdown. To make matters worse, he was simultaneously helping Weber to understand electricity. Fortunately, the interplay between electrical and magnetic forces led Riemann to a new concept of 'force', based on geometry: the same insight that led Einstein to General Relativity, decades later. Forces can be replaced by the curvature of space. Now Riemann had the new viewpoint he needed to develop his lecture.

In a somewhat desperate flurry of activity, he sorted out the foundations of modern differential geometry, beginning with the concept of a multidimensional manifold and a notion of distance defined by a metric. This is a formula for the distance between any two points that are very close together. He defined more elaborate quantities now called tensors, gave a general formula for curvature expressed as a special kind of tensor, and wrote down differential equations that determine geodesics. But he also went further, probably drawing inspiration from his work with Weber, and speculated about possible relations between differential geometry and the physical world:

> The empirical notions on which the metrical determinations of
> space are founded, the notion of a solid body and of a ray of light,
> cease to be valid for the infinitely small. We are therefore quite at
> liberty to suppose that the metric relations of space in the infinitely
> small do not conform to the hypotheses of geometry; and we ought
> in fact to suppose it, if we can thereby obtain a simpler explanation
> of phenomena.

The lecture was a triumph, even though the only person present who was likely to understand it fully was Gauss. Riemann's originality made a big impression on Gauss, who told Weber how surprised he was at its depth. The impulsive gamble had paid off.

Riemann's insights were further developed by Eugenio Beltrami, Elwin Bruno Christoffel, and the Italian school under Gregorio Ricci and Tullio Levi-Civita. Later, their work turned out to be just what Einstein needed for General Relativity. Einstein was interested in very large regions of space, whereas Riemann's vision for physics lay in the very small. Even so, it all goes back to Riemann's lecture.

Riemann's father Friedrich was a Lutheran pastor and a veteran of the Napoleonic wars. The family was poor. His mother Charlotte (*née* Ebell) died when Riemann was quite young. He had a brother and four sisters. His father educated him until the age of ten. In 1840 he started going to the local school in Hanover, entering directly into the third class. He was very shy, but his mathematical gifts were immediately apparent. The school's director allowed Riemann to read mathematics books from his own collection. When he lent the boy Legendre's 900-page text on number theory, Riemann devoured it in a week.

In 1846 he went to Göttingen University, initially to study theology, but Gauss recognised his mathematical talents and advised him to switch subjects, which (with his parents' approval) he did. Göttingen eventually became one of the best places in the world to study mathematics, but in those days, despite the presence of Gauss, its mathematical instruction was fairly ordinary. So Riemann decamped to Berlin, where he worked under the geometer Jakob Steiner, the algebraist and number theorist Dirichlet, and the number theorist and complex analyst Gotthold Eisenstein. There he learned about complex analysis and elliptic functions.

Cauchy extended calculus from real numbers to complex numbers. Complex analysis emerged when Berkeley's objections to Newton's fluxions were eventually countered by Karl Weierstrass, who formulated a rigorous definition of 'passing to a limit'. One of the hot topics in mid-1800s complex analysis was the study of elliptic functions, which among other things specify the length of an arc of an ellipse. They are a deep generalisation of trigonometric functions. Fourier exploited one basic property of these – they're periodic, repeating the same value if 2π is added to the variable. Elliptic functions have two independent complex periods, and repeat the same values on a grid of parallelograms in the complex plane. They exhibit a beautiful connection between complex analysis and symmetry groups (translations of the grid). Wiles's proof of Fermat's Last Theorem uses this idea. Elliptic functions also come up in mechanics, for example in giving an exact formula for the period of a pendulum. The simpler formula derived in school physics is an approximation for a swing through a very small angle.

Dirichlet's approach to mathematics appealed to Riemann, being much like his own. Instead of a systematic logical development, they both preferred to begin by acquiring an intuitive grasp of the problem, then sorting out the central concepts and relationships, and finally filling in the logical gaps while avoiding extensive computations as much as possible. Many of today's most successful and original mathematicians do the same. Proofs are vital to mathematics, and their logic must be impeccable – but proofs often come *after* understanding. Too much rigour, too early, can stifle a good idea. Riemann adopted this approach throughout his career. It had one big advantage: people could follow the general line of thinking without spending weeks checking complicated sums. Its disadvantage, for some at least, was the need to think conceptually, rather than just ploughing through calculations.

For his PhD, Riemann rewrote the book on complex analysis by introducing topological methods. He was led to this reformulation by a feature that every student has to grapple with: the tendency of complex functions to be *many-valued*. There are hints of this phenomenon in real analysis. For example, every nonzero positive real number has *two* square roots: one positive, the other negative. This possibility has to be borne in mind when solving algebraic equations, but it can be handled fairly easily by splitting the square root function into two separate parts: the positive square root and the negative square root.

The same ambiguity afflicts the square root of a complex number, but it's no longer entirely satisfactory to pull it apart into two distinct functions. The notions 'positive' and 'negative' have no useful meaning for complex numbers, so there's no natural way to split the two values apart. But there's a deeper issue. In the real case, if we change a positive number continuously, its positive square root also changes continuously, and so does its negative square root. Moreover, the two remain distinct. But in the complex case, continuous changes to the original number can turn one of its square roots into the other, while always moving them continuously.

The traditional way to sort this out was to allow discontinuous functions, but then you have to keep checking whether you're approaching a discontinuity. Riemann had a better idea: modify the usual complex plane to make the square root function single-valued. This is done by taking two copies of the plane, one above the other; slitting each along the positive real axis; and then joining the slits so the top plane gets joined to the bottom one as you cross the slit. The square root becomes single-valued when interpreted using this

'Riemann surface'. This is a radical approach. The idea is to stop worrying about which of the many possible values you're dealing with, and let the geometry of the Riemann surface take care of everything. And it wasn't the only innovation in the thesis. Another was to use an idea from mathematical physics, the Dirichlet principle, to prove existence of certain functions. This principle states that a function that minimises energy is a solution to a partial differential equation, Poisson's equation, which governs gravitational and electric fields. Gauss and Cauchy had already discovered that the same equation arises naturally in complex analysis in connection with differential calculus.

Riemann settled into academic life. His natural shyness made lecturing something of a trial, but he slowly adapted and began to understand how to relate to his audience. In 1857 he was appointed full professor. In the same year he published another major work on the theory of Abelian integrals, a broad generalisation of elliptic functions that provided fertile ground for his topological methods. Weierstrass had submitted an article to the Berlin Academy on the same topic, but when Riemann's paper appeared, Weierstrass was so overwhelmed by its novelty and insight that he withdrew his own work and never again published in the area. That didn't stop him pointing out a subtle error in Riemann's use of the Dirichlet principle, mind you. Riemann made heavy use of a function that made some related quantity as small as possible. It led to important results, but he hadn't given a rigorous proof that such a function actually exists. (He believed on physical grounds that it must, but this kind of reasoning lacks rigour and can go wrong.) At this stage the mathematicians split into those who wanted logical rigour, and therefore considered the gap to be serious, and those who were convinced by the physical analogies and were more interested in pushing the results further. Riemann, in the second camp, said that even though there might be a flaw in the logic, the Dirichlet principle was just the most convenient way to see what was going on, and his results were correct.

It was, in a way, a rather standard disagreement between pure mathematicians and mathematical physicists, and the same game plays out regularly today, be it the Dirac delta function or Feynman diagrams. Both sides were right, by their own standards. It makes little sense to hold up progress in physics just because some plausible and effective technique can't be justified in complete logical rigour. Equally, the absence of such justification is a smoking gun for mathematicians, hinting that something vital is missing from our understanding. Weierstrass's student Hermann Schwartz satisfied the mathematicians by finding a different proof of Riemann's results, but the physicists still preferred something more intuitive. Eventually Hilbert sorted out the existence problem by proving a version of the Dirichlet principle that is rigorous and suited to Riemann's methods. In the interim, the physicists made progress that wouldn't have happened if they'd heeded the objections of the mathematicians, and the mathematicians' efforts to justify Riemann's intuition led to a host of major results and concepts that wouldn't have been discovered if they'd sided with the physicists. Everybody won.

Manifolds and curvature had made Gauss aware of Riemann's potential and prowess, but the rest of the mathematical community got the message only after he published his research on Abelian integrals. Kummer, Karl Borchardt, and Weierstrass mentioned it when they proposed him for election to the Berlin Academy in 1859. One of the tasks facing new members was to present a report on their current work, and Riemann didn't disappoint. He'd changed tack yet again, and the report was titled 'On the number of primes less than a given magnitude'. In this work, he posed the Riemann Hypothesis, a conjecture in complex analysis related to the statistical distribution of primes. It's currently the most famous unsolved problem in the whole of mathematics.

Prime numbers are central to mathematics, but in many respects they're infuriating. They have hugely important properties, but display a remarkable absence of pattern. Looking along a list of prime numbers, in sequence, it's difficult to predict the next one (aside from everything after 2 being odd and avoiding multiples of small primes like 3, 5, 7). The primes are uniquely and unambiguously defined, yet in some respects they appear random. Statistical patterns do exist, however. Around 1793 Gauss noticed empirically that the number of primes less than any given number x is approximately $x/\log x$. He couldn't find a proof, but the conjecture became known as the prime number theorem because in those days 'theorem' was a standard term for unproved statements. Compare Fermat's Last. When a proof finally appeared, it came from a totally unexpected direction. Primes are discrete objects, arising in number theory. At the opposite end of the mathematical spectrum is complex analysis, about continuous objects, and employing totally different (geometric, analytic, topological) methods. It hardly seemed likely that there could be a connection – but there was, and mathematics has never been the same since its discovery.

The link goes back to Euler, who in 1737, in Formula Man mode, noticed that for any number s, the infinite series

$$1 + 2^{-s} + 3^{-s} + 4^{-s} + \ldots$$

is equal to the product, over all primes p, of the series

$$1 + p^{-s} + p^{-2s} + p^{-3s} + \ldots = 1/(1 - p^{-s})$$

The proof is simple, little more than a direct translation into power series language of the uniqueness of prime factorisation. Euler was thinking of this series for real numbers s, indeed mainly for integer s. But it makes sense if s is a complex number, subject to some technical issues about convergence and a trick to extend the range of numbers for which it's defined. In this context it's called the zeta function, written as $\zeta(z)$. As the power of complex analysis began to manifest itself, it was only natural to study this kind of series using the new tools, in the hope that a proof of the prime number theorem might emerge. Riemann, an expert complex analyst, was bound to get involved.

The promise of this approach first became apparent in 1848, when Pafnuty Chebyshev made progress towards a proof of the prime number theorem using the zeta function (though this name came later). Riemann made the role of this function clear in his concise but penetrating 1859 paper. He showed that the statistical properties of primes are closely related to the zeros of the zeta function, that is, the solutions z of the equation $\zeta(z) = 0$. A high point of the paper was a formula giving the exact number of primes less than a given value x as an infinite series, summed over the zeros of the zeta function. Almost as an aside, Riemann conjectured that all zeros, other than some obvious ones at negative even integers, lie on the critical line $z = {}^1/_2 + it$.

This, if true, would have significant implications; in particular, it implies that various approximate formulas involving primes are more accurate than can currently be proved. In fact, the ramifications of a proof of the Riemann Hypothesis are huge. However, no proof or disproof is known. There's some 'experimental' evidence: in 1914 Godfrey Harold Hardy proved that an infinite number of zeros lie on the critical line. Between 2001 and 2005 Sebastian Wedeniwski's program ZetaGrid verified that the first 100 billion zeros lie on the critical line. But in this area of number theory, that kind of result isn't entirely convincing, because many plausible but false conjectures first fail for absolutely gigantic numbers. The Riemann Hypothesis is part of Problem 8 in Hilbert's famous list of 23 great unsolved mathematical problems (Chapter 19), and is one of the Millennium Prize problems selected by the Clay Mathematics Institute in 2000, for which there is a prize of one million dollars for the correct solution. It's a strong contender for the biggest unsolved problem in the whole of mathematics.

Riemann proved his exact formula for primes using, among other things, Fourier analysis. The formula can be viewed as telling us that the Fourier transform of the zeros of the zeta function is the set of prime powers, plus some elementary factors. That is, the zeros of the zeta function control the irregularities of the primes. In *The Music of the Primes*, Marcus du Sautoy's title is inspired by a striking analogy. Fourier analysis decomposes a complex sound wave into its basic sinusoidal components. In the same way, the glorious symphony of the prime numbers decomposes into the individual 'notes' played by each zero of the zeta function. The loudness of each note is determined by how big the real part of the corresponding zero is. So the Riemann Hypothesis tells us that all of the zeros are equally loud.

Riemann's insights into the zeta function entitle him to be considered the musician of the primes.

16

Cardinal of the Continuum
Georg Cantor

Georg Ferdinand Ludwig Philipp Cantor
Born: St Petersburg, Russia, 3 March [OS 19 February] 1845
Died: Halle, Germany, 6 January 1918

THE CONCEPT OF INFINITY, of things that go on for ever without stopping, has intrigued human beings for millennia. Philosophers have had a field day with it. Over the last few centuries, mathematicians in particular have made extensive use of the infinite; more precisely, of a variety of different interpretations of infinity in many different contexts. Infinity isn't just a very large number. It's not really a number at all, because it's bigger than any specific number. If it were a number, it would have to be bigger than itself. Aristotle saw infinity as a process of indefinite continuation:

whichever number you've currently reached, you can always find a bigger one. Philosophers call this potential infinity.

Several Indian religions have a fascination with very big numbers. Among them is Jainism. According to the Jain mathematical text *Surya Prajnapti*, some visionary Indian mathematician stated, around 400 BC, that there are many different sizes of infinity. It sounds like mystical nonsense. If infinity is the biggest thing that can exist, how can one infinity be bigger than another? But towards the end of the nineteenth century, the German mathematician Georg Cantor developed *Mengenlehre* – set theory – and used it to argue that infinity can be actual, not just an Aristotelian process of potentiality, and that in consequence some infinities are bigger than others.

At the time, many mathematicians considered this idea to be mystical nonsense too. Cantor had to fight ongoing battles with his critics, many of whom used language that in today's world would probably result in a lawsuit. He suffered from depression, possibly exacerbated by the derision heaped upon him. But most mathematicians now accept that Cantor was right. Indeed, the distinction between the smallest infinity and any larger one is basic to many areas of applied mathematics, in particular probability theory. And set theory has become the logical foundation for the whole of mathematics. Hilbert, one of the biggest names to realise early on that Cantor's ideas were sound, said: 'No one will expel us from the paradise that Cantor has created.'

Cantor's mother, Maria Anna (*née* Böhm), was a talented musician, and his grandfather Franz Böhm had been a solo player in the Russian Imperial Orchestra. Young Georg grew up in a musical family and became an accomplished violinist. His father, also named Georg, was a wholesale agent in St Petersburg, who later joined the city's Stock Exchange. His mother was Catholic, but his father was Protestant, and Georg was brought up in that faith. Initially he had a private tutor, transferring to a primary school in the city, but St Petersburg's cold winters were bad for his father's health, so the family moved to Wiesbaden in Germany in 1856, and later to Frankfurt. Although Cantor spent the rest of his life in Germany, he later wrote that he 'never felt at ease' there, and was nostalgic for the Russia of his youth.

In Frankfurt, Cantor was a boarder at the Realschule in Darmstadt. In 1860 he graduated, being described as an unusually able student, with particular mention of his high skill in mathematics, especially trigonometry. His father wanted Cantor to become an engineer, and sent him to the Höhere Gewerbeschule in Darmstadt. But Cantor wanted to study mathematics, and pestered his father until he gave in. In 1862 he began studying mathematics at the Zürich Polytechnic. Cantor moved to the University of Berlin when his father died in 1863 and left him a substantial inheritance. There he attended lectures by Kronecker, Kummer, and Weierstrass. After a summer at Göttingen in 1866 he presented his dissertation 'On indeterminate equations of the second degree', a topic in number theory, in 1867.

He then took a position as a teacher at a girls' school, but worked on his Habilitation. After being appointed to the University of Halle, he submitted a thesis in number theory, and Habilitation was granted. Eduard Heine, a prominent mathematician at Halle, suggested that Cantor should change fields and tackle a famous unsolved problem about Fourier series: prove that the representation of a function in this form is unique. Dirichlet, Rudolf Lipschitz, Riemann, and Heine himself had all tried to prove this result, but failed. Cantor solved it within a year. For a time he continued working on trigonometric series, and his researches led him into areas that we now recognise as prototype set theory. The reason is that many properties of Fourier series rest on delicate features of the function being represented, such as the structure of the set of points at which it is discontinuous. Cantor couldn't make progress in these areas without coming face to face with complicated issues about infinite sets of real numbers.

Research into the foundations of mathematics were on the rise, and after centuries of informal treatment of the 'real' numbers as infinite decimals, mathematicians were starting to wonder what it all meant. For example, there's no way to write down the infinite decimal expansion of π. All we can do is give rules for how to find it. In 1872 one of Cantor's papers about trigonometric series introduced a novel method for defining a real number as the limit of a convergent sequence of rational numbers. In the same year Dedekind published a famous paper, in which he defined a real number in terms of a 'section' dividing the rational numbers into two disjoint subsets, such that the members of one subset are all less than any member of the other. In it, he cited Cantor's paper. These two approaches – convergent sequences of rationals or Dedekind sections – are both standard in courses on the foundations of mathematics and the construction of the set of real numbers from the rationals.

By 1873 Cantor had embarked on the research that qualifies him as a significant figure of the highest order: set theory and transfinite (his term for infinite) numbers. Set theory has since become an essential part of any mathematics course, because it gives a convenient and versatile language in which to describe the subject. Informally, a set is any collection of objects; they could be numbers, triangles, Riemann surfaces, permutations, whatever. Sets can be combined in various ways. For example, the union of two sets is what you get by combining them into one set, and the intersection is what they have in common. Using sets, we can define basic concepts such as functions and relations. We can construct systems of numbers such as the integers, rationals, reals, and complex numbers from simpler constituents, making heavy use of the empty set, which has no members.

Transfinite numbers are a way to extend the notion of 'how many members?' to infinite sets. Cantor stumbled across this idea in 1873 when he proved that the rational numbers are countable; that is, they can be placed in one-to-one correspondence with the natural numbers 1, 2, 3, ... (I'll explain the ideas and terminology shortly.) If there's only one size of infinity, this result would be obvious, but he soon found a proof that the real numbers are *not* countable. It was published in 1874, a year of great personal importance for Cantor because he married Vally Guttmann – a marriage that would lead to six children.

Seeking a still larger infinity than that of the reals, Cantor thought about the set of all points in the unit square. Surely the square, with its two dimensions, has more points than the real line? Writing to Dedekind, Cantor expressed his opinion:

> Can a surface (say a square that includes the boundary) be uniquely referred to a line (say a straight line segment that includes the end points) so that for every point on the surface there is a corresponding point of the line and, conversely, for every point of the line there is a corresponding point of the surface? I think that answering this question would be no easy job, despite the fact that the answer seems so clearly to be 'no' that proof appears almost unnecessary.

Soon, however, he found that the answer wasn't as obvious as it seemed. ('Proof appears unnecessary' to a mathematician is like a red rag to a bull, and he should have seen it coming.) In 1877 he proved that such a correspondence does, in fact, exist. 'I see it, but I don't believe it!' Cantor wrote. But when he submitted his paper to the prestigious *Journal für die reine und angewandte Mathematik* (Journal for Pure and Applied Mathematics), Leopold Kronecker – a brilliant but ultra-conservative mathematician and a leading light of that period – was unconvinced, and only the intervention of Dedekind led to the work's acceptance and publication. Cantor, with some justification, never submitted another paper to that journal. Instead, between 1879 and 1884, he sent the bulk of his development of set theory and transfinite numbers to the *Mathematische Annalen* (Mathematical Annals), probably facilitated by Felix Klein.

Before continuing Cantor's story, we need to understand the revolutionary nature of his ideas, and what they were about. It would be too confusing to present them in the terminology of the period, so I'll apply some modern hindsight to extract a few basic ideas.

In his 1638 *Discourses Relating to Two New Sciences*, Galileo raised a basic issue – somewhat paradoxical – about infinity. The book is presented as a discussion between Salviati, Simplicio, and Sagredo. Salviati always wins, Simplicio doesn't stand a chance, while Sagredo's job is to keep the discussion moving. Salviati observes that it's possible to match counting numbers to squares, so that each number corresponds to a unique square, and each square to a unique number. Just match each number with its square:

1	2	3	4	5	6	7	8	9	10	11	12
$\updownarrow$	$\updownarrow$	$\updownarrow$	$\updownarrow$	$\updownarrow$	$\updownarrow$	$\updownarrow$	$\updownarrow$	$\updownarrow$	$\updownarrow$	$\updownarrow$	$\updownarrow$
1	4	9	16	25	36	49	64	81	100	121	144

With finite numbers, if two sets of objects can be matched in this manner, they must contain the same number of members. If everyone seated at a table has their own knife and fork, and just one of each, then the number of knives equals that of forks, and both equal the number of people. So, even though squares form a rather 'thin' subset of all numbers, it seems that there are exactly as many squares as numbers. Salviati concludes: 'We can infer only that the totality of all numbers is infinite, and the attributes "equal", "greater", and "less", are not applicable to infinite, but only to finite, quantities.'

Cantor realised that the situation isn't quite that bleak. He used this kind of matching (which he called a one-to-one correspondence) to define 'same number of members' for sets, be they finite or infinite. This can be done, interestingly enough, without knowing what the number actually is. Indeed, we've just done so for the knives and forks. So logically, 'same number' is prior to 'number'. There's nothing strange about this: we can see when two people are equally tall without knowing their exact heights, for example.

The way to introduce actual numbers is to specify a standard set, and say that anything that matches it has that set as its cardinal – a fancy word for 'number of elements'. The obvious choice for an infinite set is the set of all natural numbers, which defines a transfinite cardinal that Cantor dubbed 'aleph-Null'. Here aleph is the first letter of the Hebrew alphabet, and Null is German for 'zero'. In symbols, it looks like this: $\aleph_0$. By definition, any set that matches the natural numbers has cardinal $\aleph_0$. Salviati proved that the set of squares also has cardinal $\aleph_0$.

This seems paradoxical because there are clearly numbers that are not squares – indeed, 'most' numbers aren't squares. We can resolve the paradox by accepting that removing some elements from an infinite set need not make its cardinal smaller. The whole need not be greater than the part, as far as cardinals are concerned. However, we don't have to follow Salviati and reject the whole idea of comparison: we get sensible results if we assume that the whole is greater than *or equal to* the part. After all, the whole point about infinity as a concept is that it doesn't always behave like finite numbers. The big question is how far we can get, and what we can salvage.

Cantor's next big discovery was that the rational numbers (let's work with the positive ones for simplicity) also have cardinal $\aleph_0$. They can be matched to the natural numbers like this:

$$
\begin{array}{ccccccccccc}
1/1 & 1/2 & 2/1 & 1/3 & 3/1 & 1/4 & 2/3 & 3/2 & 4/1 & 1/5 & 5/1 \\
\updownarrow & \updownarrow & \updownarrow & \updownarrow & \updownarrow & \updownarrow & \updownarrow & \updownarrow & \updownarrow & \updownarrow & \updownarrow \\
1 & 2 & 3 & 4 & 5 & 6 & 7 & 8 & 9 & 10 & 11
\end{array}
$$

To get the top row, we order rationals differently from their numerical order. Define the complexity of a rational number to be the sum of the numerator and the denominator. Consider only rationals where these have no common factor, to avoid including the same number twice. For example, 2/3 and 4/6 are the same rational; we choose only the first form. First split the rationals into classes, ordered by complexity. Each such class is finite. Then, within each class, order the fractions according to their numerators. So the class with complexity 5 is ordered like this:

$$1/4 \quad 2/3 \quad 3/2 \quad 4/1$$

It's easy to prove that every positive rational occurs once and once only. The natural number that matches it is its position in the resulting ordered list.

Up to this point, it might be that $\aleph_0$ is just a fancy symbol for infinity, and all infinities are equal. The next discovery explodes that possibility. The set of real numbers can't be matched to the natural numbers.

Cantor's first proof of 1874 was aimed at a problem in number theory, the existence of transcendental numbers. An algebraic number is one that satisfies some polynomial equation with integer coefficients, such as $\sqrt{2}$, which satisfies $x^2–2 = 0$. If a number isn't algebraic, it's called transcendental. No such equation was known for e or π, and these were thought to be transcendental, a conjecture that turned out to be correct. Liouville proved the existence of a transcendental number in 1844, but his example was very artificial. Cantor proved that 'most' real numbers are transcendental, by showing that the set of algebraic numbers has cardinal $\aleph_0$, but the set of reals has a larger cardinal. His proof involves assuming the reals are countable, and constructing a sequence of nested intervals that omit every real number in turn. The intersection of those intervals (which can be proved not to be empty) must contain a real number, but whichever one that is, it's already been excluded.

In 1891 he found a more elementary proof, the famous diagonal argument. Assume (for a contradiction) that the real numbers (say between 0 and 1 for simplicity) are countable. Then the counting numbers can be matched to these reals. In decimal notation, any matching of this kind takes the form

1     $0 \cdot a_1 a_2 a_3 a_4 \ldots$

2     $0 \cdot b_1 b_2 b_3 b_4 \ldots$

3     $0 \cdot c_1 c_2 c_3 c_4 \ldots$

4     $0 \cdot d_1 d_2 d_3 d_4 \ldots$

. . .   . . .

By assumption, every real number occurs somewhere in the list. Now we construct one that doesn't. Define successive decimal places $x_1, x_2, x_3, \ldots$ of a real number x as follows:

If $a_1 = 0$ let $x_1 = 1$, otherwise let $x_1 = 0$.

If $b_2 = 0$ let $x_2 = 1$, otherwise let $x_2 = 0$.

If $c_3 = 0$ let $x_3 = 1$, otherwise let $x_3 = 0$.

If $d_4 = 0$ let $x_4 = 1$, otherwise let $x_4 = 0$.

Continue this process indefinitely, making x_n either 0 or 1, so that it differs from the nth decimal digit of the real number corresponding to n.

By construction, x differs from every number on the list. It differs from the first number in its first digit, from the second number in its second digit; in general, it differs from the nth number in its nth digit, so it's different from the nth number, no matter what value n has. However, we assumed that the list exists, and every real number appears on it. This is a contradiction, and what it contradicts is the assumption that such a list exists. Therefore no such list exists, and the set of real numbers is uncountable.

A similar idea underpins Cantor's discovery, which he found hard to believe, that the plane has the same cardinal as the real line. A point in the plane has coordinates (x, y) where x and y are real numbers. For simplicity, restrict to the unit square; then x and y have decimal expansions

$$x = 0 \cdot x_1 x_2 x_3 x_4 \ldots$$

$$y = 0 \cdot y_1 y_2 y_3 y_4 \ldots$$

Match this pair to a point on the line whose coordinate is those of x and y interleaved, like this:

$$0 \cdot x_1 y_1 x_2 y_2 x_3 y_3 \ldots$$

Since we can recover x and y by selecting only successive digits in odd- or even-numbered locations, this defines a one-to-one correspondence between the unit square and the unit interval on the line. It's easy to beef this up to the entire plane and the entire line. (A few technicalities need to be taken care of, which I've suppressed, to do with the lack of uniqueness of the decimal representation of a number.)

There was one question that Cantor was unable to decide, either way. Is there a transfinite cardinal strictly between $\aleph_0$ and the cardinal of the real numbers? Cantor thought not, because he couldn't find one, despite trying a lot of plausible candidates. This conjecture became known as the Continuum Hypothesis. We'll see how it fared in Chapter 22.

✳

For a decade from 1874, Cantor threw his efforts into set theory, discovered the importance of one-to-one correspondences in the foundations of the number system, and devised his extension of counting principles to transfinite numbers. His work was so original that many of his contemporaries were unable to accept it or believe it had value. His mathematical career was blighted by Kronecker, who found his revolutionary ideas philosophically distasteful. 'God made the integers, all else is the work of Man,' Kronecker said.

Cantor rather set himself up as a philosophical target by stating unequivocally that set theory was about actual infinity, not Aristotelian potential infinity. This was a slight overstatement because it's 'actual' infinity only in a conceptual sense. In mathematics, it's usually possible to pass from a description that appears to involve actual infinity to one that looks purely potential. However, this translation process often looks contrived: Cantor was correct that the natural way to think about his work is to view infinity as a completed whole, not as a process that, while finite at any stage, can be continued indefinitely. The philosopher Ludwig Wittgenstein was a vocal critic. He was especially scathing about the diagonal argument, and even when Cantor had died he was still complaining about 'the pernicious idioms of set theory'. But the main reason he kept complaining was that mathematicians increasingly sided with Cantor, and none of them paid much attention to Wittgenstein. This must have been especially galling since he was particularly interested in the philosophy of mathematics, but then mathematicians don't take kindly to philosophers who insist they're doing it all wrong. Set theory *worked*, and most mathematicians are pragmatic, even about foundational issues.

Cantor was religious, and struggled to reconcile his mathematics with his beliefs. The nature of the infinite was still heavily bound up with religion, because the Christian God was considered infinite, and was held to be the unique actual infinity. Kronecker's remark about the integers wasn't a metaphor. Then along comes Cantor, claiming actual infinities in mathematics ... Well, you can see what would happen. Cantor struck back, though, saying: 'The transfinite species are just as much at the disposal of the intentions of the Creator ... as are the finite numbers.' This was a clever argument, because denying it would be to claim God had limitations, which was heretical. Cantor even wrote to Pope Leo XIII about it all, and sent him some mathematical articles. God knows what the Pope thought about them.

✳

Others understood what Cantor was doing. Hilbert recognised the importance of Cantor's work, and praised it. But as he grew older, Cantor felt that set theory had not made the impact he'd hoped for. In 1899 he had an attack of depression. He soon recovered, but he lost confidence, telling Gösta Mittag-Leffler 'I don't know when I shall return to the continuation of my scientific work. At the moment I can do absolutely nothing with it.' To combat his depression he went for a holiday in the Harz mountains, and attempted a reconciliation with his academic enemy, Kronecker. Kronecker responded positively, but the atmosphere between them remained tense.

Cantor's mathematics was a worry, too: he was unhappy that he couldn't prove his Continuum Hypothesis; thought he'd proved it was false, but quickly found a mistake; then thought he'd proved it was true, but again found a mistake. At this point Mittag-Leffler asked Cantor to withdraw a paper from *Acta Mathematica* even though it had reached proof stage – not because it was wrong, but because it was 'one hundred years too soon'. Cantor joked about this, but he was very hurt. He stopped writing to Mittag-Leffler, took no further interest in the journal, and pretty much gave up on set theory.

His depression tended to express itself in two ways. One was an increased interest in the philosophical implications of set theory. The other was a conviction that the works of Shakespeare were actually written by Francis Bacon. This bee in his bonnet led him to make a serious study of Elizabethan literature, and by 1896 he was publishing pamphlets about this pet theory. Then, in quick succession, his mother, younger brother, and youngest son died. He showed increasing signs of mental instability, and in 1911, when the

University of St Andrews in Scotland invited him as a distinguished guest to celebrations of the university's 500th anniversary, he spent much of the time talking about Bacon and Shakespeare. Depression became a constant companion. He spent some time in hospital for the condition, and in 1918 he died in a sanatorium from a heart attack.

The irony is that Mittag-Leffler was essentially right when he told Cantor he was a century ahead of his time, though not perhaps in the sense he intended. Although Cantor's ideas slowly gained ground, the most significant impact of set theory on mathematics had to wait until the 1950s and 60s, when the abstract approach to mathematics promoted by the group calling itself Nicolas Bourbaki came into full flower. Bourbaki's influence on mathematical education has (thankfully) waned, but its insistence that mathematical concepts should be defined precisely, in as much generality as possible, still holds sway. And the basis for precision and generality is the viewpoint afforded by Cantor's beloved sets. Today, every area of mathematics, pure and applied, is firmly based in the formalism of set theory. Not just philosophically, but practically. Without the language of sets, mathematicians now find it impossible even to specify what they are talking about.

The verdict of posterity is that, yes, there are philosophical issues with set theory and transfinite numbers, but these are no worse than the very similar philosophical issues with Kronecker's beloved integers. Those, too, are the work of man, and the work of man is usually flawed. Ironically, we now define them using ... set theory. And we see Cantor as one of the true originals in mathematics. If he hadn't invented set theory, someone would eventually have done so, but it could well have taken decades before anyone else came along with his unique combination of power, depth, and insight.

17

The First Great Lady
Sofia Kovalevskaia

Sofia Vasilyevna Kovalevskaia (*née* Korvin-Krukovskaya) or
Sophie/Sonya Kowalevski
Born: Moscow, Russia, 15 January 1850
Died: Stockholm, Sweden, 10 February 1891

FROM EARLIEST CHILDHOOD young Sofa, as the family affectionately called her, had a burning desire to understand whatever took her fancy. Her interest in mathematics was kindled at the age of eleven; remarkably, the cause was the nursery wallpaper. Her father Vasily Korvin-Krukovsky was lieutenant-general of artillery in the Imperial Russian Army, and her mother Yelizaveta (*née* Shubert) was from a family of high standing in the Russian nobility. The wallpaper comes into the story because the family owned a country estate at Palabino, near St Petersburg. On moving to Palabino the family had the whole house redecorated, but failed to buy enough wallpaper for the nursery. As a substitute, they used pages from an old textbook, which just happened to be Ostrogradskii's lecture course on differential and integral calculus. In her autobiography *Memories of Childhood* Sofia remembered spending hours staring at the walls, trying to figure out the meaning of the arcane symbols that covered them. She quickly memorised the formulas, but later recalled that 'at the time I was studying it I could not understand it at all'.

She already had form in this kind of self-education. The fashion at the time was not to teach reading to young children, but Sofia had been desperate to read. At the age of six she had taught herself by memorising the shapes of letters in newspapers and then pestering an adult to tell her what they meant. She showed off her new ability to her father, who, though incredulous at first – he thought she'd just memorised a few sentences – was soon convinced, and was immensely proud of her initiative and intelligence.

When Sofia's bedroom wallpaper triggered a similarly self-propelled interest in mathematics, her family, remarkably forward-thinking for the period, did nothing to discourage it, even though many among their social circle would not have deemed mathematics to be a fit subject for a young lady. Circumstances conspired to allow her to pursue her passion. Mathematics had been one of her father's favourite subjects, and Sofia was his favourite daughter. Her mother's father Fedor Fedorovich Shubert had been a military topographer, and *his* father Fedor Ivanovich Shubert had been a leading astronomer and a member of the Academy of Sciences. So mathematical blood (to use that period's image of heredity) flowed in Sofia's veins. Moreover, her family had long been immersed in the mathematical subculture, which may well have been a more important influence.

Beginning with the basics, the general made sure that Sofia's tutors instructed her in arithmetic. But when he eagerly asked his daughter how she liked it, her initial response was distinctly lukewarm: it wasn't calculus. Her view changed when she finally realised that without the basics, she would never progress to those fascinating equations on the wallpaper. Not only did she go on to master calculus; she progressed to the frontiers of mathematical research, making discoveries that amazed the period's leading mathematicians. She worked on partial differential equations, mechanics, and the diffraction of light by crystals. Her mathematical publications number only ten, and one is a translation into Swedish of one of the others, but their quality is outstanding. She was penetrating, original, and technically proficient. The prominent American mathematician Mark Kac described her as the 'first great lady of mathematics'. She was arguably the greatest female scientist of her time, eclipsed only by Marie Curie a few decades later.

Sofia was born in Moscow in 1850. She had an elder sister, Anna, known to the family as Aniuta, whom she adored; later she was blessed with a younger brother, Fedor. Her uncle Pyotr Vasilievich Krukovsky had a strong interest in mathematics, and often talked to her about it, long before she could possibly understand what he was saying.

In 1853, when Sofia was three years old, Russia became embroiled in the Crimean War. The conflict was ostensibly about the rights of Christian minorities in the Holy Land, but France and the United Kingdom were determined to stop Russia from taking over areas of the declining Ottoman Empire. By 1856 an alliance of France, the United Kingdom, Sardinia, and the Ottomans had defeated Russia after the siege of Sevastopol. This humiliation brought about massive public discontent in Russia. Peasants and liberals revolted against an oppressive system, which they increasingly viewed as corrupt and incompetent. The government fought back with censorship and repression by the tsarist secret police. Many nobles owned vast country estates, but they seldom spent much time there, preferring St Petersburg's political importance and social delights. Prudence now dictated that even those with liberal leanings should spend more time in the country and pay more attention to the grievances of their workforce. So in 1858 General Korvin-Krukovsky told his wife that it had become their duty to relocate to their estate.

At first Sofia and her elder sister Aniuta were left to their own devices, exploring the countryside and generally getting into scrapes. But after they tried to eat some unsuitable berries, and were ill for days, their father hired a Polish tutor Iosif Malevich and a strict English governess Margarita Smith, who the girls disliked intensely. Malevich taught Sofia the basics of a young woman's education, including arithmetic, but her uncle Petr inducted her into some of the mysteries of more advanced mathematics – topics such as squaring the circle (constructing a square of area equal to that of a given circle, which is actually impossible with the traditional geometric instruments of ruler and compasses) and asymptotes (lines that a curve approaches indefinitely close to without ever reaching them). These concepts fired her imagination and left her wanting more.

Eventually Miss Smith resigned, and peace reigned in the Korvin-Krukovsky household. In 1864 Aniuta sent two stories that she'd written to Fedor and Mikhail Dostoievski, which were published in their journal *Epokha*. Aniuta began a secret correspondence with Fedor, and after her father objected and then relented, Fedor Dostoievski became part of the family's circle. Sofia joined the social whirl, meeting other prominent figures as well. For a time she developed a schoolgirl crush on Dostoievski. When Fedor proposed marriage to Aniuta, Sofia was outraged, even more so when her sister turned him down.

At about this time she became absorbed in the mathematical mysteries of her bedroom wallpaper, and one strand of her future life was set. A neighbour, Nikolai Tyrtov, was a physics professor at the Petersburg Naval Academy, and he brought her a copy of his introductory textbook on physics. Not knowing any trigonometry, she struggled until she found a more intuitive geometric approximation – essentially the clasical use of a chord of a circle. Tyrtov, excited by this demonstration of her abilities, urged the general to let her study higher mathematics.

At that time, Russian women were not allowed to go to university, but they could study abroad with written permission from father or husband. So Sofia contracted a 'fictitious marriage' with Vladimir Kovalevskii, a young palaeontology student. This ploy, a marriage of convenience with no genuine relationship, was fairly common among educated young Russian women as a way to gain some freedom. To Sofia's chagrin, her father suggested a delay. In typically headstrong manner she bided her time until the house was full of distinguished dinner guests; then she sneaked out, leaving a note saying she had gone to Vladimir's lodgings unchaperoned, and would stay there until they were allowed to marry. To avoid social meltdown, the general duly presented his daughter and her fiancé to the guests. Sofia's plan was to get married, then dump Vladimir and go her own way, but Vladimir became infatuated with his future wife and her social circle, and had no wish for them to separate. They married in 1868, when Sofia was 18, and she became Sofia Kovalevskaia.

Like many young Russians at that time, Kovalevskaia's political views were nihilist. That is, she rejected any convention that lacked rational support, such as government and the law. Vladimir Lenin, quoting the radical writer Dmitri Pisarev, captured the attitude, an extreme form of social Darwinism thrown back in the faces of the rich and powerful who often justified their privileges in much the same way: 'Break, beat up everything, beat and destroy! Everything that's being broken is rubbish and has no right to life! What survives is good.' When the newlyweds arrived in St Petersburg their apartment soon became the social hub for like-minded nihilists.

In 1869 they left Russia, initially for Vienna. Vladimir's publishing business had collapsed and he was fleeing creditors; both of them were also seeking a more intellectual atmosphere. Vladimir set his sights on geology and palaeontology. Kovalevskaia – to her surprise – was permitted to take physics lectures at the university, but in the absence of any equally accommodating mathematicians the couple moved to Heidelberg. At first the university authorities gave her the usual runaround, apparently under the impression she was a widow and bemused when told she was married, but eventually it was agreed that she was free to attend lectures provided the professor had no objection. Soon she was spending twenty hours a week in lectures, by mathematicians such as Leo Königsberger and

Paul DuBois-Reymond, the chemical physicist Gustav Kirchhoff, and physiologist Hermann Helmholtz.

She also pestered the misogynist chemist Wilhelm Bunsen to allow her and her friend Iulia Lermontova to work in his laboratory, where he had previously sworn no woman – especially a Russian – would ever set foot. 'Now *that woman* has made me eat my words,' he complained to Weierstrass, and spread scandalous rumours in revenge. His colleagues, in contrast, were enthusiastic about their talented female student, and the newspapers carried occasional articles about her. Kovalevskaia refused to let the attention go to her head, and concentrated on her studies.

The Kovalevskiis travelled to England, France, Germany, and Italy. Vladimir met Charles Darwin and Thomas Huxley, with whom he was already acquainted. Through such contacts, Kovalevskaia was able to meet the novelist George Eliot socially. In her journal for 5 October 1869 Eliot wrote: 'On Sunday, an interesting Russian pair came to see us – M. and Mme. Kovalevskii: she, a pretty creature, with charming modest voice and speech, who is studying mathematics ... at Heidelberg; he, amiable and intelligent, studying the concrete sciences apparently – especially geology.' The philosopher and social Darwinist Herbert Spencer was also present, and boorishly proclaimed the intellectual inferiority of women. Kovalevskaia argued with him for three hours, and Eliot wrote that she had 'defended our common cause well and bravely'.

In 1870 Kovalevskaia moved to Berlin, hoping to study under Weierstrass. Hearing rumours that he disapproved of education for women, she wore a bonnet more suited to an older woman, which hid her face. Weierstrass was surprised when she asked to study with him, but replied politely, giving her some problems to solve and bring back. A week later she returned, having solved them all, often by original methods. Weierstrass later said she had 'the gift of intuitive genius'. The university senate refused her permission to study officially, so Weierstrass offered her private lessons. They began a correspondence that continued until her death.

Aniuta was by now living in Paris with Victor Jaclard, a young

Marxist. In 1871 the National Guard declared the Paris Commune, a radical socialist government that briefly ruled the city. Lenin said it was 'the first attempt by the proletarian revolution to smash the bourgeois state machine'. The state machine had no wish to be smashed. When Sofia heard that Jaclard might be arrested for his political activities, the Kovalevskiis headed for Paris. As the Versailles government began to bombard the Commune, Sofia and Aniuta nursed the wounded. The Kovalevskiis returned to Berlin, but when Paris fell and Jaclard was arrested, they went back to help Aniuta, getting her safely to London. There, Karl Marx provided more help. General Korvin Krukovsky and his wife went to Paris intending to get Jaclard set free. They couldn't secure an official release, but it was causally mentioned that Jaclard was being moved to another prison. As the prisoners were being taken through the crowds, a woman grabbed him by the arm and hauled him away. Some believe it was Aniuta (except she was in London at the time), some Kovalevskaia, others Jaclard's sister; some think it was Vladimir in disguise. Jaclard escaped, Vladimir gave him his passport, and he fled to Switzerland. From then on, even when immersed in her mathematics, Kovalevskaia involved herself in political and social movements.

Back in Berlin, she dived into her studies with enthusiasm. Her research was going well, but her marriage was not. The couple quarrelled incessantly, and Vladimir was muttering darkly about a divorce. By 1874 Kovalevskaia had written three research articles, all of doctoral quality. The most important was the first; Charles Hermite called it 'the first significant result in the general theory of partial differential equations'. The second was on the dynamics of Saturn's rings, and the third was a technical paper about the simplification of integrals.

A partial differential equation relates the rates of change of some quantity with respect to several distinct variables. For instance, Fourier's heat equation relates changes in temperature with respect to space – along the rod – to how its value at each specific location changes with respect to time. His trick for solving the equation using trigonometric series relies on a special feature: the equation is linear, so solutions can be added to each other to give further solutions. Kovalevskaia's paper of 1875 proves the existence of solutions for *nonlinear* partial differential equations, provided they satisfy some technical conditions. It extended Cauchy's results from 1842, and a combined version is now called the Cauchy–Kovalevskaia theorem.

Her paper on Saturn's rings was written while she worked with Weierstrass, but the topic didn't interest him and she did the research alone. She studied the dynamics of revolving rings of liquid, which Laplace had proposed as a model for the rings of Saturn. She analysed the stability of the rings in this model, showing that they couldn't be ellipses as Laplace had thought, but must be egg-shaped, fat at one end and thinner at the other. The paper is interesting for its methods, and would have been even more so if it had contained the necessary proofs, but it soon became known that the rings were made of innumerable discrete particles, so the underlying fluid model was questionable. As Kovalevskaia wrote: 'Due to Maxwell's research it has become doubtful whether Laplace's view of the structure of the rings of Saturn is acceptable.'

Now came the perennial problem of academic politics. The papers had to be presented to a university for a doctorate, and it had to be one of the rare institutions that was willing to award one to a woman. Weierstrass approached Göttingen, which sometimes awarded doctorates to foreigners without making them undergo the usual formal oral examination, which would be carried out in German. Kovalevskaia obtained the degree of doctor of philosophy in mathematics *summa cum laude* (with distinction), becoming the first woman after Maria Agnesi in Renaissance Italy to gain a PhD in mathematics, and one of precious few to get a scientific doctorate.

Kovalevskaia was now a fully fledged mathematician.

In 1874 the Kovalevskiis went back to Russia, first to the family home in Palabino, and from there to St Petersburg in search of academic positions. They failed to secure any job offers. Kovalevskaia's German degree counted for nothing: she would need a Russian one. However, as a woman, she wasn't permitted to sit the exam. Frustrated, the Kovalevskiis went into business to make some money, a decision that quickly proved disastrous. Kovalevskaia's father died in 1875, leaving her a legacy of 30,000 roubles, which would have allowed them a frugal living, if invested wisely. Instead, they put the money into a property scheme. Initially this seemed to be a success, and the Kovalevskiis moved to a new house, with a garden, orchard, and cow. (Having your own cow was *de rigueur* among wealthy middle-class Russians.) The couple had a daughter, also named Sofia. Vladimir put more money into a radical newspaper, eventually losing 20,000 roubles when it folded. Months later the property scheme collapsed. Vladimir had used speculative future profits to buy land, and when his creditors called in his debts, the property empire turned out to be fantasy.

In 1878 Kovalevskaia renewed contact with Weierstrass and followed his advice to tackle the refraction of light by a crystal. In 1879 she lectured to the Sixth Congress of Natural Scientists on her earlier research on Abelian integrals. In 1881 she and her daughter arrived in Berlin, where Weierstrass had found them an apartment. Vladimir's finances went from bad to worse, and the couple's possessions were sold to pay off part of his debt. In 1883, suffering from sudden mood swings and likely to face prosecution for his role in a financial swindle, he committed suicide by drinking a bottle of chloroform. A guilt-ridden Kovalevskaia stopped eating for five days, then fainted. Force-fed by her doctor, she regained consciousness and threw herself into her work, completing her theory of refraction in a crystal. She returned to Moscow to set Vladimir's affairs in order, and presented her refraction research to the Seventh Congress of Natural Scientists.

Her husband's death removed a major obstacle between Kovalevskaia and an academic post, because a widow was more acceptable than an independent or married woman. Previously, she had met the leading Swedish mathematician Gösta Mittag-Leffler through his sister Anna Carlotta Edgren-Leffler, a revolutionary, actress, novelist, and playwright. Their friendship lasted until Kovalevskaia's death. Mittag-Leffler, impressed by her research on Abelian integrals, secured a position for her at Stockholm University – temporary and provisional, but a genuine academic post nonetheless. Kovalevskaia became the only woman to hold such a position in the whole of Europe. She arrived in Stockholm late in

1883. She knew the job would be challenging, and she'd have to battle against prejudice, but one progressive newspaper described her as 'a princess of science', which was encouraging. Though she did remark that a salary would be even better.

Kovalevskaia's literary ambitions blossomed, and she and Edgren-Leffler coauthored two plays: *The Struggle for Happiness* and *How It Might Have Been*. She also attacked a major classical problem in mechanics: the rotation of a rigid body about a fixed point. Here she made a totally unexpected discovery – a new type of solution now called the Kovalevskaia top. Some academic-political horse-trading converted her unpaid position into an extraordinary professorship, which might be made permanent after five years. Now she had enough to live on – barely – and began paying off some of her late husband's debts. She became something of a local celebrity, which persuaded the University of Berlin to allow her to attend lectures in any Prussian university. She travelled back to Russia, then to Berlin, and back to Sweden. She joined the editorial board of the journal *Acta Mathematica*, another female first.

Wheels were moving; Hermite had persuaded the Paris Academy's Prix Bordin to set a problem tailored to her interests, and there was little doubt among the inner circle that Kovalevskaia would win. In 1888 she duly won the prize for her work on the rotation of a solid body. As her reputation as a major research mathematician grew, the old barriers were starting to come down. In 1889 she was appointed professor ordinarius at Stockholm University, a tenured lifetime post. She was the first woman to hold such a position at a northern European university. After much lobbying on her behalf, she was granted a Chair in the Russian Academy of Sciences. First the committee voted to change the rules to allow women to be admitted; three days later they elected her.

Kovalevskaia wrote several non-mathematical works, including *A Russian Childhood*, her plays with Anna Carlotta, and a partly autobiographical novel, *Nihilist Girl* (1890). She died of influenza in 1891.

✳

Kovalevskaia's unexpected discovery of a new solution to the problem of a rotating rigid body was a major contribution to mechanics, which is about how particles and bodies move under the action of forces. Typical examples are the swing of a pendulum, the spinning of a top, and the orbital motion of a planet round the Sun. As we saw in Chapter 7, mechanics really took off in 1687 when Newton published his laws of motion. The second law is especially important because it tells us how a body moves under the influence of known forces: mass times acceleration equals force. This law specifies the body's position indirectly in terms of the *rate of change of the rate of change* of position, making it a 'second order' differential equation.

If we're lucky, we can solve the equation, obtaining a formula for the position of the body at any given time. If so, the equation is integrable. Much early work in mechanics boils down to finding systems modelled by integrable equations. But even for very simple systems, this can be hard. A pendulum is one of the simplest mechanical systems there is, and it does turn out to be integrable; even so, an exact formula involves elliptic functions.

To begin with, integrable cases were discovered by intelligent trial and error. As mathematicians gained experience, they started to pin down some general principles. The most important of these are known as conservation laws, because they specify quantities that are conserved – don't change – during the motion. The most familiar is energy. In the absence of friction, the total energy of a mechanical system always remains the same. Others are linear and angular momentum. If there are enough conserved quantities, they can be used to deduce the solution, and the system is integrable. For historical reasons, the integrable cases for the motion of a rigid body are referred to as 'tops'.

Before Kovalevskaia, two integrable tops were known. One is the Euler top, a rigid body not subject to external twisting forces (torques). The other is the Lagrange top, which spins about its axis on a flat horizontal surface with gravity acting vertically. Lagrange discovered that this system is integrable if the top has rotational symmetry. The key in both cases is to consider the top's moments of inertia, which tell us how much torque (twisting force) is needed to accelerate its angular motion about a given axis by a given amount. Every rigid body has three special moments of inertia, said to be principal. Every mathematician versed in mechanics knew about the Euler and Lagrange tops. They also knew – or thought they did – that these were the only integrable cases. So Kovalevskaia's discovery of a third type was, to say the least, a shock. Moreover, it doesn't rely on symmetry, which mathematicians were starting to get accustomed to and realised helped to solve equations; instead, her new solution exploited mysterious features of a top with one principal moment of inertia half the size of the other two. We now know there are no other integrable cases to be found.

Systems that are not integrable can be studied by other means, such as numerical approximations. Often, they exhibit deterministic chaos: irregular behaviour resulting from non-random laws. But even today, physicists, engineers, and mathematicians take a serious interest in integrable systems: they're easier to understand, providing rare islands of regularity in an ocean of chaos, and their exceptional nature makes them special, hence worthy of detailed study. The Kovalevskaia top has become a classic of mathematical physics.

18

Ideas Rose in Crowds
Henri Poincaré

Jules Henri Poincaré
Born: Nancy, Lorraine, France, 29 April 1854
Died: Paris, France, 17 July 1912

ARCHIMEDES GOT IDEAS in the bath. Henri Poincaré got them stepping on a bus.

Poincaré was one of the most inventive and original mathematicians of his time. He also wrote several bestselling popular science books based on lectures given to the Parisian Société de Psychologie. He took an interest in the thought processes of mathematicians, placing particular emphasis on the subconscious mind. In *Science and Method* he relates an example from his own experience:

> For fifteen days I strove to prove that there could not be any functions like those I have since called Fuchsian functions. I was

then very ignorant; every day I seated myself at my table, stayed an hour or two, tried a great number of combinations and reached no results. One evening, contrary to my custom, I drank black coffee and could not sleep. Ideas rose in crowds; I felt them collide until pairs interlocked, so to speak, making a stable combination. By the next morning I had established the existence of a class of Fuchsian functions, those which come from the hypergeometric series; I had only to write out the results, which took but a few hours.

He then goes into some detail about his own experiences, first pointing out that, to paraphrase, you don't need to know what the technical terms in the story mean. Just consider them placeholders for some advanced mathematical topic.

I wanted to represent these functions by a quotient of two series; this idea was perfectly conscious and deliberate, the analogy with elliptic functions guided me. I asked myself what properties these series must have if they existed, and I succeeded without difficulty in forming the series I have called theta-Fuchsian. Just at this time I left Caen, where I was then living, to go on a geological excursion under the auspices of the school of mines. The changes of travel made me forget my mathematical work. Having reached Coutances, we entered an omnibus to go some place or other. At the moment when I put my foot on the step the idea came to me, without anything in my former thoughts seeming to have paved the way for it, that the transformations I had used to define the Fuchsian functions were identical with those of non-Euclidean geometry. I did not verify the idea; I should not have had time, as, upon taking my seat in the omnibus, I went on with a conversation already commenced, but I felt a perfect certainty. On my return to Caen, for conscience' sake I verified the result at my leisure.

The story continues with two further moments of sudden illumination.

Introspecting on this and other discoveries, Poincaré distinguished three phases of mathematical discovery: preparation, incubation, and illumination. That is: do enough conscious work to immerse yourself in the problem and get stuck; wait while the subconscious mulls it over; then the little light bulb in your head goes off, the celebrated 'aha!' moment.

It's still one of the best insights we have into the workings of a great mathematical mind.

✳

Henri Poincaré was born in Nancy, France. His father Léon was professor of medicine at the University of Nancy, and his mother was Eugénie (*née* Launois). His cousin Raymond Poincaré became prime minister, and was president of the French Republic during the first world war. Henri suffered from diphtheria when quite young, and his mother gave him special tuition at home until he had recovered. He went to the Lycée in Nancy, spending eleven years there. He came top in every subject and was absolutely formidable in mathematics. His teacher called him a 'monster of mathematics' and he won national prizes. He had an excellent memory and could visualise complicated shapes in three dimensions, which helped compensate for eyesight so poor that he could hardly see the blackboard, let alone what was written on it.

In 1870 the Franco-Prussian war was in full swing, and Poincaré served in the ambulance corps with his father. The war ended in 1871, and in 1873 he attended the École Polytechnique in Paris, graduating in 1875. Then he switched to the École des Mines, studying mining engineering as well as more mathematics. He got a degree in mining engineering in 1879. It was a busy year. He became a mine inspector in the Corps des Mines in the Vesoul region, and carried out an official investigation into an accident at Magny in which eighteen miners were killed. He also pursued his doctorate under Hermite, working on difference equations, analogues of differential equations in which time changes in discrete steps rather than continuously. He realised the potential of such equations as models of many bodies moving under gravity, such as the solar system, anticipating future developments along those lines, which grew in importance when computers became powerful enough to carry out the huge numbers of calculations required.

After obtaining his doctorate, he got a job as junior lecturer in mathematics at the University of Caen and met his future wife Louise Poulin d'Andesi. They married in 1881, and had four children – three girls and a boy. By 1881 Poincaré had secured a far more prestigious job at the University of Paris, where he matured into one of the leading mathematicians of his age. He was highly intuitive, and his best ideas often arrived when he was thinking about something else – as his story about the bus exemplifies. He wrote several bestselling popular science books: *Science and Hypothesis* (1901), *The Value of Science* (1905), and *Science and Method* (1908). And he ranged over most of the mathematics of his day, including complex function theory, differential equations, non-Euclidean geometry, topology – which he virtually founded – and applications of mathematics to areas as diverse as electricity, elasticity, optics, thermodynamics, relativity, quantum theory, celestial mechanics, and cosmology.

Topology, you'll recall, is 'rubber-sheet geometry'. Euclid's geometry is built around properties preserved by rigid motions, such as lengths, angles, and areas. Topology throws all this away, seeking properties preserved by continuous transformations, which can bend, stretch, compress, and twist. Among them are connectedness (one piece or two?), knottedness, and having one or more holes. The subject may sound nebulous, but continuity is fundamental, perhaps even more so than symmetry. In the twentieth century topology became one of the three pillars of pure mathematics, the others being algebra and analysis.

That it did so owes a large debt to Poincaré, who went beyond rubber sheets to rubber spaces, so to speak. The metaphor of a sheet is a two-dimensional concept. Ignoring any surrounding space – Gauss's viewpoint – it takes only two numbers to specify a point on a sheet, or, more formally, a surface. The classical topologists, among them Gauss's student Johann Listing, managed to understand the topology of surfaces in considerable detail. In particular, they classified them; that is, they listed all of the possible shapes. To do this, they exploited an ingenious method to construct a surface from a flat polygon (and its interior).

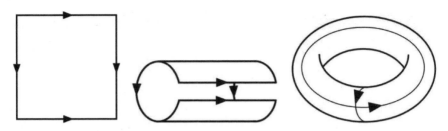

If opposite edges of a square are glued together, the result is a torus. But the result can be imagined and studied using just the square and the gluing rules, without actually bending the square.

A simple and important example of a surface is the torus. When embedded in three-dimensional space, this is shaped like an American doughnut, with a hole through the middle. A mathematical torus is defined to be the surface of the doughnut – no dough, just the boundary between dough and the surrounding air. Conceptually, this shape can be defined without any dough or air. Start with a square, and add rules saying that corresponding points on opposite edges are identical. *If* you were to bend the square round to glue corresponding edges together, you'd get a torus shape. But you can study everything on a flat square, provided you remember the rules. Many computer games 'bend' the rectangular screen by implementing the gluing rules graphically, so an alien monster that disappears off the left-hand edge reappears at the right. No one with any sense would try *physically* to bend the screen to achieve this effect. This object goes by the oxymoronic name of the 'flat torus'. It's flat because its local geometry is the same as that of a flat square. It's a torus because its global topology is that of ... well, a torus.

Listing and others showed that any closed surface of finite extent can be obtained by conceptually gluing the edges of a suitable polygon. It usually has more than four sides, and the gluing rules can be complicated. From this it can be proved that every orientable surface – having two separate sides, unlike the famous Möbius band – is a k-torus. That is, it is a surface like a torus but with k holes in it, for $k = 0, 1, 2, 3, \ldots$ If $k = 0$ we get a sphere, if $k = 1$ we get the usual torus, and if $k \geqslant 2$ we get something more complicated. There's an analogous classification of non-orientable surfaces too, but let's not go into that.

The 2-torus and 3-torus.

Poincaré wanted to generalise topology to spaces of dimension greater than two, and the obvious first step was to go to three dimensions. Here Gauss's intrinsic view of geometry is vital, because it makes little sense to try to embed a complicated topological space in ordinary three-dimensional Euclidean space. It's like trying to embed a torus in the plane, without the trick of identifying edges. It won't fit.

To see that interesting three-dimensional topological spaces – three-dimensional manifolds, or 3-manifolds – are possible, we generalise the trick that Listing used. For example, the flat three-dimensional torus is made by taking a solid cube (to get something three-dimensional we need the interior, not just the six square faces) and conceptually gluing opposite faces together. Now a solid alien could vanish through one face and reappear at the opposite one, as if those two faces are opposite sides of a Stargate-style doorway and the alien has just passed through it.

More generally, we can take a polyhedron and glue faces together according to some list of rules. This recipe leads to lots of topologically different 3-manifolds, but it no longer gives them all. (That's not obvious, but it's true.) In fact, classifying the topological types of manifolds with three or more dimensions is essentially impossible; there are too many topologically distinct shapes. But with enough effort, some general patterns can be dug out. In this connection an absolutely basic question goes back to Poincaré; it's known as the Poincaré Conjecture. Really it would be better named the Poincaré Mistake, as we'll shortly see, but let's be charitable. In 1904 Poincaré discovered that something he'd tacitly been assuming was obvious wasn't even true, and he asked whether it could be fixed up by starting from stronger hypotheses. He couldn't sort it out, remarking that 'this question would lead us too far astray', and he left it as a teaser for future generations.

To understand the conjecture, we begin with an analogous question in the simpler context of surfaces. How can you distinguish the sphere from all the other k-tori? Poincaré noticed that a simple topological feature does the trick. If you draw a loop – a curve whose ends join together – on a sphere, it can be deformed continuously, always staying on the sphere, until it's all scrunched up at a single point. With no holes to get in the way, you can just keep shrinking the loop until it all piles up at that point. On a k-torus with one or more holes $(k > 0)$, however, a loop that winds through a hole can't be shrunk like that. It has to stay threaded through the hole.

The jargon for 'every loop deforms to a point' is 'homotopy sphere'. We've just sketched a proof that, for surfaces, every homotopy sphere is topologically the same as a genuine sphere. This characterises a sphere by a simple topological property. A hypothetical ant, living on a surface, could in principle figure out whether it's on a sphere by dragging loops of string around and trying to pile them up at a single point. Poincaré assumed that the same kind of thing characterises a 3-sphere, which is a 3-manifold analogous to a spherical surface. It's not just a solid ball. A ball has a boundary, the 3-sphere doesn't. You can think of it as a solid ball whose surface is scrunched up into a single point – just as a disc turns into a sphere, topologically, when you gather all the boundary points together. Think of a bag with a string around the top. When you pull the string tight, the boundary scrunches up, and the bag has the same topology as a sphere.

Now do this when you have one extra dimension to play with.

The conjecture came about because Poincaré was thinking about another topological property, known as homology. This is less intuitive than deforming loops, but it's closely related. There's a sense in which loops threaded through distinct holes in a k-torus constitute independent ways *not* to be deformable to a point. Homology captures this idea without reference to holes, which are an interpretation of the outcome in terms that appeal to our visual sense. The notion of a hole is a bit misleading, because a hole isn't part of the surface: it's a place where the surface is absent. In two dimensions, thanks to the classification theorem, a sphere can be characterised by its homology properties (no holes).

In one of his early papers, Poincaré assumed that the same statement is true for three dimensions. It seemed so obvious that he didn't bother to prove it. But then he discovered a space that has the same homology as the 3-sphere, but is topologically distinct from the 3-sphere. To make it, glue opposite faces of a solid dodecahedron together, much like making a flat three-dimensional torus from a solid cube. To prove that this 'dodecahedral space' is not topologically equivalent to a 3-sphere, Poincaré invented homotopy – what happens to a loop when you deform it. Unlike the 3-sphere, his dodecahedral space contains loops that *don't* deform continuously to a point. So then he asked whether this extra property does characterise the 3-sphere. It was a question, not really a conjecture, because he didn't express an explicit opinion. However, it's clear he expected the answer to be yes, so calling it a conjecture isn't too unfair.

The Poincaré Conjecture turned out to be hard. Very hard. If you're a topologist, used to the terminology and ways of thinking, the question is simple and natural. It ought to have a natural answer with a simple proof. Apparently not. But the ideas that led Poincaré to it sparked an explosion of research into topological spaces and properties like homology and homotopy, which, if you're lucky, can distinguish them. The Poincaré Conjecture was finally proved in 2002 by Grigori Perelman, using new methods inspired in part by General Relativity.

For Poincaré, topology wasn't just an intellectual game. He applied it to physics. The traditional method for analysing a dynamical system is to write down its differential equation and then solve it. Unfortunately, this method seldom gives an exact answer, so for centuries mathematicians used approximate methods. Until computers became widely available, the approximations took the form of infinite series, of which only the first few terms would actually be used; computers made numerical approximations practical as well. In 1881, Poincaré developed an entirely new way of thinking about differential equations in 'Memoir on curves defined by a differential equation'. This paper founded the qualitative theory of differential equations, which seeks to deduce properties of the solutions of a differential equation without writing down formulas or series for them, or calculating them numerically. Instead, it exploits general topological features of the phase portrait – the collections of all solutions, viewed as a unified geometric object.

A solution of a differential equation describes how the variables change with the passage of time. A solution can be visualised by plotting these variables as coordinates. As time passes, the coordinates change, so the point that they represent moves along a curve, the solution trajectory. The possible combinations of variables determine a multidimensional space, with one dimension per variable, called phase space or state space. If solutions exist for all initial conditions, which is commonly the case, every point in phase space lies on some trajectory. So phase space breaks up into a family of curves, the phase portrait. The curves fit together like smoothly combed fur made of long hairs, except near a steady state of the equation, where the solution remains constant for all time and the hair reduces to a point. Steady states are easy to find, and provide the beginnings of a 'skeleton' of the phase portrait: a diagram of its main distinctive features.

As described so far, we have to know the solutions, or numerical approximations to them, to draw the phase portrait. Poincaré discovered that some properties of the solutions can be detected topologically. For example, if the system has a periodic solution – one that repeats the same sequence of states over and over again – the trajectory is a loop, and the solution just keeps going round and round like a pet hamster in a wheel. Topologically, any loop can be deformed into a circle, so the problem simplifies to topological properties of circles. The presence of a loop can sometimes be detected by considering a Poincaré section. This is a surface that cuts across a bundle of trajectories. Given any point on the section, we follow its trajectory until (if ever) it hits the section again. This determines a map from the surface to itself, the Poincaré map or 'first return' map. If the section crosses a periodic trajectory, the corresponding point returns to the same location. That is, it's a fixed point of the Poincaré map.

In particular, suppose that the section is a disc, a ball, or a higher-dimensional analogue, and that we can show that the image of the section under the Poincaré map lies inside the same section. Then we can invoke a general topological result known as the

Brouwer fixed-point theorem to conclude that a fixed point must exist; that is, the differential equation has a periodic solution passing through that section. Poincaré introduced a variety of techniques along these lines, and stated a general conjecture about the long-term behaviour of trajectories for differential equations in two variables. Namely, the trajectory can converge to a point, a closed loop, or a heteroclinic cycle – a loop formed by trajectories that link a finite number of fixed points together. Ivar Bendixson proved this conjecture in 1901, and the result is known as the Poincaré–Bendixson theorem.

Poincaré's realisation that topological methods potentially offer deep insights into the solutions of differential equations, even when there's no formula for those solutions, lies behind today's approach to nonlinear dynamics, with applications across the scientific board. It led him to another epic discovery: chaos, now one of the big triumphs of topological dynamics. The context was the motion of several bodies under Newtonian gravity – the many-body problem.

Johannes Kepler deduced from observations of Mars that the orbit of a single planet round the Sun is an ellipse. Newton explained this geometric fact in terms of his law of gravitation: *any two bodies in the universe attract each other with a force proportional to their masses and inversely proportional to the square of the distance between them.* In principle, Newton's law predicts the motion of any number of mutually gravitating bodies, such as the planets of the solar system. Unfortunately the law of gravity doesn't prescribe the movement directly: it provides a differential equation whose *solution* gives the positions of the bodies at any instant of time. Newton found that, for two bodies, this equation can be solved, and the result is Kepler's ellipse. But for three or more bodies, no tidy solution of this kind seemed feasible, and mathematicians working in celestial mechanics resorted to special tricks and approximations.

The year 1889 saw the sixtieth birthday of Oscar II, King of Sweden and Norway, which at the time formed a single kingdom. In celebration, the king offered a prize for a solution of the many-body problem, a topic proposed by Mittag-Leffler. The answer was to be given not as a simple formula, which almost certainly didn't exist, but as a convergent infinite series. The problem can then be solved to any desired degree of accuracy by calculating enough terms of the series.

Poincaré decided to have a go, and won the prize, even though his memoir didn't solve the full problem. He considered only three bodies, and assumed two were of equal mass orbiting each other at diametrically opposite points of a circle, and the third was so light that it had no effect on the two more massive bodies. His results presented evidence that, in some circumstances, no solution of the specified kind exists. The system can sometimes behave in a highly irregular manner, so that its geometry looks as though someone has accidentally dropped a loosely wound ball of string on the ground. He described his key geometric insight, about how two important curves defining the dynamics cross each other:

> When one tries to depict the figure formed by these two curves and their infinity of intersections, each of which corresponds to a doubly asymptotic solution, these intersections form a kind of net, web or infinitely tight mesh ... One is struck by the complexity of this figure that I am not even attempting to draw.

We now understand that Poincaré had discovered the first important example of dynamical chaos: the existence of solutions to deterministic equations, so irregular that some aspects of them seem to be random. But at the time, this result – though intriguing – seemed more like a dead end.

Until recently, what I've just written was the official story. But in the 1990s the mathematical historian June Barrow-Green was visiting the Mittag-Leffler Institute in Sweden. She came across a printed copy of a different version of Poincaré's memoir – and it didn't mention the possibility of highly irregular orbits. It turned out that this was the version Poincaré submitted, but after the winner was announced, he noticed an error. Almost the entire print run was destroyed, and a corrected version was quickly printed at Poincaré's expense. One copy of the original, however, survived in the institute's archives.[10]

✳

Poincaré might have given the appearance of the stereotypic impractical academic, but he retained his mining connections throughout his life, and from 1881 to 1885 he directed the development of the northern railway as an engineer at the Ministry of Public Services. In 1893 he was made chief engineer of the Corps de Mines, and in 1910 he was promoted to inspector general. At the University of Paris he occupied chairs in many subjects: mechanics, mathematical physics, probability, and astronomy. His election to the Academy of Sciences came when he was only 32, in 1887, two years before he won King Oscar's prize, and he eventually became its president in 1906. In 1893 he worked for the Bureau des Longitudes, trying to set up a unified system of time throughout the world, suggesting that the world should be divided up into time zones.

He came very close to beating Einstein to Special Relativity, showing in 1905 that Maxwell's equations for electromagnetism are invariant under what we now call the Lorentz group of transformations, which implies that the speed of light must be constant in a moving frame. Perhaps the main point he missed, but Einstein spotted, was that physics really is like that. He also proposed the notion of a gravitational wave, propagating at the speed of light, in the flat spacetime of Special Relativity. The LIGO experiment detected such waves in 2016, but by then the context had shifted to the curved spacetimes of General Relativity.

Poincaré died from an embolism after cancer surgery in 1912, and was buried in his family's vault at the cemetery of Montparnasse. His mathematical reputation continued to grow, as others developed the ideas that he first proposed. Today he is considered to be one of the great originals of the subject, and one of the last to range over almost the entire mathematical landscape of his day. His mathematical legacy remains alive and kicking.

19

We Must Know, We Shall Know
David Hilbert

David Hilbert
Born: Wehlau, near Königsberg, Prussia (now Kaliningrad, Russia), 23 January 1862
Died: Göttingen, Germany, 14 February 1943

A GERMAN PROFESSOR who reached the age of sixty-eight was obliged to retire. When David Hilbert passed this milestone in 1930, many public events marked the official end of an outstanding academic career. He lectured on his first big result, the existence of a finite basis for invariants. Motorists found themselves driving along the newly named Hilbertstrasse. When his wife remarked 'What a nice idea!' Hilbert replied 'The idea, no – but the execution is nice.'

Most pleasing of all was being made an honorary citizen of Königsberg, the city near which he had been born. The honour was to be conferred at a meeting of the Society of German Scientists and Physicians, and Hilbert had to deliver an acceptance speech. He decided it must be widely accessible, and since Immanuel Kant had been born in Königsberg, something with a philosophical aspect would be appropriate. It must also sum up his life's work. He settled on 'Natural knowledge and logic'. Hilbert had form in such activities, often delivering lectures in a Saturday morning series intended for everyone in the university. Relativity, infinity, the principles of mathematics ... he did his best to make them accessible to anyone interested. Now he focused all his efforts on a lecture that would trump them all.

'Understanding nature and life is our noblest task,' he began. He went on to compare and contrast two ways of understanding the world: thought and observation. The two are linked by the laws of nature, to be deduced from observations and developed by pure logic. It was a view that would have appealed to Kant, which was ironic because Hilbert wasn't a big fan of Kant. This wasn't the occasion to say so, and on this particular issue there was no disagreement, but Hilbert couldn't resist one dig, a suggestion that Kant had overestimated the importance of *a priori* knowledge, not obtained through experience. Geometry was a good example: there was no reason to assume that space was necessarily Euclidean, as Kant had argued. Remove the anthropomorphic dross, however, and true *a priori* concepts remain, namely, the generalities of mathematics. 'Our entire present culture, insofar as it is concerned with the intellectual understanding and conquest of nature, rests on mathematics!' he declaimed. And he ended by defending pure mathematics, often criticised for lack of practical relevance: 'Pure number theory is that part of mathematics for which *up to now* [my italics] no application has ever been found ... The glory of the human spirit is the sole aim of all science!'

So successful was the talk that Hilbert was persuaded to repeat it for the local radio station, and the recording survives. He emphasised that problems previously thought impossible – such as finding the chemical composition of a star – had yielded to new ways of thinking. 'There is no such thing as an insoluble problem,' he said. The final words of the talk were 'We must know. We shall know.' Then, just as the technician stopped the tape, Hilbert laughed.

At the time, Hilbert was deep into a massive programme to set the whole of mathematics on logical foundations, and his words were a statement of confidence that his programme would succeed. Already, much progress had been made. A few stubborn cases still had to be sorted out. When they were polished off, Hilbert wouldn't just have a logical basis for all mathematics: he'd be able to prove that his axioms are logically consistent.

It didn't work out the way he'd hoped.

Hilbert came from a family of lawyers. His grandfather was a judge and privy councillor, his father Otto a county judge. His mother Maria (*née* Erdtmann) was a Königsberg merchant's daughter. Her passions were philosophy, astronomy, and prime numbers, and it looks as though her enthusiasms rubbed off on her son. A sister Elsie arrived when David was six. His mother taught him at home until he entered school at the late age of eight. The school specialised in the classics, offering little in the way of mathematics and no science whatsoever. Rote learning was the order of the day, and Hilbert did poorly at anything requiring memorising structureless lists of facts. He describes himself as having been 'dull and silly'. One subject was a glorious exception. His school report reads: 'For mathematics he always showed a very lively interest and a penetrating understanding: he mastered all the material taught in the school in a very pleasing manner and was able to apply it with sureness and ingenuity.'

In 1880 Hilbert began studying for a degree at the University of Königsberg, specialising in mathematics. He took courses at Heidelberg under Lazarus Fuchs; back in Königsberg he studied under Heinrich Weber, Ferdinand von Lindemann, and Adolf Hurwitz. He became close friends with Hurwitz and with Hermann Minkowski, a fellow student. Throughout his life he corresponded with Minkowski. Lindemann, who was shortly to become famous for proving that π does not satisfy any algebraic equation with integer coefficients, became Hilbert's thesis advisor. He suggested that Hilbert should work in invariant theory, following the trail blazed by Boole and extended by Cayley, Sylvester, and Paul Gordan. Their methods were computational, and Hilbert's proficiency in these horrible calculations impressed his friend Minkowski, who wrote 'I rejoiced over all the processes which the poor invariants had to pass through'. In 1885 Hilbert was awarded his doctorate, after giving a public lecture on physics and philosophy.

At that time the leading authority on invariant theory was Gordan, and the big unsolved question was to prove the existence, for any number of variables and any degree of equation, of a finite basis. That is, a finite number of invariants, such that all other invariants are combinations of them. List the basis, and in effect you've got the lot. For two-variable quadratics, the basis consists solely of the discriminant. Finiteness had been proved in many cases, always by calculating all of the invariants and then extracting a basis. By this method, Gordan had proved the most general known theorem of this kind.

The entire area was turned upside down in 1888, when Hilbert published a short paper proving that a finite basis always exists, *without calculating any invariants whatsoever.* In fact, he proved that any suitable collection of algebraic expressions always has a finite basis – be it composed of invariants or not. This wasn't the kind of answer that Gordan had been expecting, and when Hilbert submitted the work to *Mathematische Annalen*, Gordan rejected it. 'This is not mathematics,' he said. 'This is theology.' Hilbert complained to the editor Klein, refusing to change the paper unless some 'definite and irrefutable objection against my reasoning is raised'. Klein agreed to publish the paper in its original form. I suspect he understood the proof better than Gordan, who was out of his depth when computational ability was replaced by conceptual thinking.

A few years later Hilbert extended his results and submitted another paper. Klein accepted it, describing it as 'the most important work on general algebra that the *Annalen* has ever published'. As far as Hilbert was concerned, he had now achieved everything he had set out to do in that area. 'I shall definitely leave the field of invariants,' he wrote to Minkowski. And he did.

✳

Having polished off invariant theory – the subject pretty much died once Hilbert had demolished it, only to be revived many years later in a still more general context, and with renewed interest in computations as well as concepts – Hilbert found a new area to work in. In 1893 he embarked on a new project, the *Zahlbericht* (Number Report). The German Mathematical Society had asked him to survey a major area within number theory, to do with algebraic numbers. These are complex numbers that satisfy a polynomial equation with rational (equivalently, integer) coefficients. An example is $\sqrt{2}$, which satisfies $x^2 - 2 = 0$; another is the imaginary number i, which satisfies $x^2 + 1 = 0$. As remarked in Chapter 16, a complex number that is not algebraic is called transcendental (page 170); examples include π and e, though this property is hard to prove and for a long time was an open problem. Charles Hermite proved e transcendental in 1873, and Lindemann dealt with π in 1882.

The main role played by algebraic numbers was in number theory. Euler had tacitly used some of their properties, for example when proving Fermat's Last Theorem for cubes, but it was Gauss who began a systematic study. When trying to generalise his law of quadratic reciprocity to higher powers than the square, he discovered a beautiful extension to fourth powers, based on algebraic numbers of the form $a + ib$ where a and b are integers. This system of 'Gaussian integers' has many special features, and in particular it has its own analogue of prime numbers, complete with a unique factorisation theorem. Gauss also made use of algebraic numbers related to roots of unity in his construction of the regular heptadecagon.

In Chapter 6, in connection with Fermat's Last Theorem, we discussed Kummer's use of algebraic numbers and his notion of ideal numbers (page 60). Dedekind simplified this idea by reformulating it in terms of special *sets* of algebraic numbers, which he called ideals. After Kummer, algebraic number theory took off, aided and abetted by Galois's theory of equations and the growing development of abstract algebra (Chapter 20). The phrase 'algebraic number theory' has two interpretations: an algebraic approach to number theory, or the theory of algebraic numbers. Both meanings were now converging on to the same thing, and this is what the German Mathematical Society wanted Hilbert to sort out. Characteristically, he went much further. He asked a perennial question among mathematicians when faced with a large body of impressive but disorganised results: 'Yes, but what's it *really* about?' This led him to formulate and prove many new theorems.

Throughout the preparation of the *Zahlbericht*, Hilbert got extensive feedback from Minkowski – sometimes too extensive, so that at times Hilbert began to despair of ever finishing it to his friend's satisfaction, but eventually the report was published. It formulated and proved general analogues of quadratic reciprocity, providing the basis of what is now called class field theory, still a flourishing, though highly technical, framework for algebraic number theory. The preface of the *Zahlbericht* states:

> Thus we see how far arithmetic, the Queen of mathematics, has conquered broad areas of algebra and function theory to become their leader ... The conclusion, if I am not mistaken, is that above all the modern development of pure mathematics takes place under the banner of number.

We might not go quite so far today, but at the time the claim was justifiable.

Hilbert would spend five to ten years in one area, polish off the big problems, and then depart for pastures new – sometimes forgetting completely that he'd ever studied that topic. He once remarked that he did mathematics because you can always work something out again if you forget it. A mathematician's mathematician to the core, he had now 'done' algebraic number theory. He moved on. His students, who had been bombarded with lectures on algebraic numbers for years, were startled to find that next year's topic was the elements of geometry. Hilbert was going back to Euclid.

As always, Hilbert had his reasons, and again the key question was: 'Yes, but what's it *really* about?' Euclid's answer would have been 'space', which is why he illustrated his theorems with geometric drawings. Hilbert, however, was far more interested in the logical structure of the axioms for geometry and how they led to theorems that were often far from obvious. He was also dissatisfied with Euclid's list of axioms, because the use of pictures had led Euclid to make assumptions that he had *not* stated explicitly.

A simple example is 'a straight line passing through a point that lies inside a circle must meet the circle'. This looks obvious in a picture – but it's not a logical consequence of Euclid's axioms. Hilbert realised that Euclid's axioms were incomplete, and set out to remedy the deficiency. Euclid defined a point as 'that which has no part', and a straight line to be 'a line which lies evenly with the points on itself'. Hilbert considered these statements to be meaningless. What matters, he argued, is how these concepts behave, not some mental image of what they are. 'One must be able to say at all times – instead of points, straight lines, and planes – tables, chairs, and beer mugs,' Hilbert told his colleagues. In particular, pictures were out.

This project was of course related to the deeper question, by then well understood, of non-Euclidean geometries and the parallel axiom (Chapter 11). Hilbert was trying to establish the basic principles for axiomatic treatments of mathematical topics. These included consistency (not leading to a logical contradiction) and independence (no axiom is a consequence of the others). Other desirable qualities were completeness (nothing vital is missing) and simplicity (when possible). Euclidean geometry was a test case. Consistency was easy: you can *model* Euclid's geometry using algebra, applied to (x, y) coordinates in the plane. That is, you can start with ordinary numbers and construct a mathematical system obeying all of Euclid's axioms. Then the axioms can't be self-contradictory, because proof by contradiction would then show that the constructed model *does not exist*. There's one potential flaw in that argument, however, and Hilbert was aware of it early on. It assumes that the standard system of numbers is itself non-contradictory; that arithmetic is consistent, which is what mathematicians mean by 'exists'. Obvious though this may seem, no one had actually proved it. Later, Hilbert tried to eliminate this gap, but it came back to haunt him.

The outcome was a small, concise, and elegant book,

Foundations of Geometry, published in 1899. It developed Euclidean geometry from 21 explicit axioms. Three years later Eliakim Moore and Robert Moore (no relation) proved that one of them can be deduced from the others, so only 20 are actually needed. Hilbert started with six primitive notions: 'point', 'line', 'plane', and the relations 'between', 'lies on', and 'congruent'. Eight axioms govern incidence relations among points and lines, such as 'any two distinct points lie on a line'. Four (which Euclid's pictures led him to assume without making them explicit) govern the order of points along a line. Six more sort out congruence (of line segments and triangles; 'congruent' basically means 'same shape and size'). Next came Euclid's parallel axiom, which by then every competent mathematician knew had to be included. Finally, there were two subtle axioms of continuity, ensuring that the points of a line are modelled on the real numbers (and not, say, the 'rationals, where lines that seem to meet in a diagram may fail to do so at a rational point).

The main value of Hilbert's book wasn't in teaching – Euclid wasn't exactly in vogue any more – but in unleashing a flurry of activity on the logical foundations of mathematics. American mathematicians in particular were at the forefront of this wave, from which emerged a logico-mathematical hybrid, metamathematics. This is, in a sense, mathematics applied to itself; more properly, to its own logical structure. A mathematical proof can be viewed not just as a process that leads to new mathematics, but as a mathematical object in its own right. Indeed, it's this deep self-referential aspect that sowed the seeds of the destruction of Hilbert's dream. In November of the same year came the bombshell, a paper by a young logician named Kurt Gödel (Chapter 22). It contained proofs of two devastating theorems. First, if mathematics is consistent, this can never be proved. Second, there exist statements in mathematics for which neither a proof nor a disproof exists. Mathematics is inherently incomplete, its logical consistency cannot be determined, and some problems are truly impossible to solve.

Hilbert is reported as being 'very angry' when he first learned of Gödel's work.

✳

No account of Hilbert's influence is complete without mentioning the Hilbert Problems, a list of 23 major open questions and areas in mathematics, which he presented in a talk at the Second International Congress of Mathematicians in Paris in 1923. They set the scene for a substantial proportion of mathematical research in the twentieth century. They include finding a consistency proof for mathematics, a rather vague request for an axiomatic treatment of physics, questions about transcendental numbers, the Riemann Hypothesis, the most general reciprocity law in any number field, an algorithm to determine when a Diophantine equation has a solution, and various technical issues in geometry, algebra, and analysis. Ten have been completely solved, three remain unsolved, a few are too vague to recognise what a solution would look like, and two don't have solutions at all, in a very strong sense.

Mathematics after Hilbert didn't just consist of people trying to solve his 23 problems, but they exerted a considerable, and largely beneficial, influence on the development of mathematics over the next half-century. If you wanted to make your mark among your mathematical fellows, solving a Hilbert problem was a good way to go.

Hilbert's interest in mathematical physics became stronger as he aged, a common phenomenon among mathematicians who begin their careers in the 'pure' camp and gradually drift towards applications. By 1909 he was working on integral equations, leading to the notion of a Hilbert space, now fundamental to quantum mechanics. He also came close to discovering Einstein's equations for General Relativity in a 1915 paper, published five days before Einstein's announcement, stating a variational principle that implies the Einstein equation. However, he failed to write down the equation itself.

Hilbert was usually a genial soul, lavish with praise for a good performance, but he could be brutal if anyone uttered meaningless platitudes or lied to him. In seminars, if a student was labouring some point that Hilbert felt was straightforward, he would say 'But that is completely simple!' and a wise student would promptly move on. In the 1920s Hilbert ran a Mathematics Club, with weekly meetings, open to anyone. Many well-known mathematicians gave talks, instructed to present 'only the raisins out of the cake'. If the calculations became difficult, Hilbert would interrupt with 'we are not here to check that the sign is right'.

As time passed, he became less tolerant. Alexander Ostrowksi remarked that once, when a visitor gave an excellent talk on a really important and beautiful piece of research, Hilbert's only question was a sour: 'What's it good for?' When Norbert Wiener, a brilliant American who coined the term 'cybernetics', talked to the club, everyone went out for supper, as was the custom. Hilbert started talking about past lectures at the Club, saying that the quality had generally declined over the years. In his day, he said, people really thought about content and presentation, but nowadays young people usually gave poor talks. 'Recently it has been especially bad,' he said. 'But now, this afternoon, there was an exception –'

Wiener prepared for the compliment.

'This afternoon's talk was the worst there ever has been!'

In 1933 the Nazis were rooting out the Jews among the Göttingen academics and dismissing them. One was Hermann Weyl, one of the great mathematical physicists, who had been appointed Hilbert's successor on his retirement in 1930. Others were Emmy Noether (Chapter 20), the number theorist Edmund Landau, and Paul Bernays, Hilbert's collaborator on mathematical logic. By 1943 virtually the entire department of mathematics had been replaced by people more acceptable to the Nazi administration, and was a pale shadow of its former glorious self. That year, Hilbert died.

He'd seen it all coming. A few years earlier Bernhard Rust, the minister of education, had asked Hilbert whether Göttingen's Mathematics Institute had suffered because of the departure of the Jews. It was a stupid question, because most of the faculty had either been Jews or their spouses. Hilbert's answer was direct and blunt:

'Suffered? It doesn't exist any longer, does it?'

20

Overthrowing Academic Order
Emmy Noether

Amalie Emmy Noether
Born: Erlangen, Germany, 23 March 1882
Died: Bryn Mawr PA, USA, 14 April 1935

IN 1913 EMMY NOETHER, a female mathematician of great renown, was in Vienna giving a course of lectures, and she visted Franz Mertens, a mathematician who worked in many fields but is best known for his contributions to number theory. One of Mertens's grandsons later wrote down his recollections of the visit:

> Although a woman, [she] seemed to me like a Catholic chaplain from a rural parish – dressed in a black, almost ankle-length and

rather nondescript, coat, a man's hat on her short hair ... and with
a shoulder bag carried crosswise like those of the railway
conductors of the imperial period, she was rather an odd figure.

Two years later this unassuming person was responsible for one of
the great discoveries of mathematical physics: a fundamental link
between symmetries and conservation laws. From that point on,
symmetries of the laws of nature have played a central role in
physics. Today they underpin the 'standard model' of subatomic
particles in quantum theory, which is virtually impossible to describe
without appealing to symmetry.

Noether was a leading figure in the development of abstract
algebra, in which calculations with many different types of numbers
or formulas were organised in terms of the algebraic laws that these
systems obey. Perhaps more than any other mathematician, the 'odd
figure' seen by Mertens's grandson was responsible for the change
that marks the borderline between the neoclassical period of the
nineteenth and early twentieth century, with its emphasis on special
structures and formulas, and the modern period from about 1920
onwards, with emphasis on generality, abstraction, and conceptual
thought. She was the inspiration behind the subsequent Bourbakiste
movement, which originated in the joint efforts of a group of young,
mainly French, mathematicians, who aimed to make mathematics
precise and general. Perhaps *too* general, in some eyes at least, but
there you go.

Emmy Noether was born into a Jewish family in the Bavarian town
of Erlangen. Her father Max was a notable mathematician who
worked in algebraic geometry and algebraic function theory. He was
highly talented but, compared to the greats of his era, a bit
specialised. His family was well off, owning a flourishing wholesale
hardware company. This background undoubtedly influenced
Emmy's attitudes to life and to mathematics. Initially, she planned to
become a teacher, and obtained the necessary qualifications to teach
French and English. But, perhaps not so surprisingly, she was bitten

by the mathematics bug and studied at the University of Erlangen, where her father worked.

Two years earlier the university senate had declared that mixed-sex education would 'overthrow all academic order', and there were only two women students out of 986. She was allowed to audit classes but not to participate fully, and had to get permission from individual professors to attend their lectures. But in 1904 the rules changed, allowing women to matriculate on the same basis as men. Noether duly did so in 1904, and moved to Gauss's old stamping ground, the University of Göttingen, to do a PhD in invariant theory, supervised by the eminent Gordan. Her thesis calculations were extraordinarily complex, culminating in a list of 331 'covariants' for fourth-degree forms in three variables. The normally indefatigable Gordan had given up on this gigantic calculation forty years earlier. Noether's methods were fairly old-fashioned, paying little or no attention to Hilbert's innovations. In 1907 she received the PhD degree *summa cum laude*.

If Noether had been a man, she would naturally have progressed to the next stage in securing a permanent academic post. But women were not allowed to proceed with Habilitation, so she worked unpaid at Erlangen for seven years. She helped her father, who by then was disabled, and continued her own research. A formative experience, which diverted her towards more abstract methods, was a series of discussions with Ernst Fischer, who drew her attention to Hilbert's new methods and advised her to use them. This she did, spectacularly, and its effects are visible throughout her subsequent career.

Mathematics was starting to open up to women, and Noether was admitted to several major mathematical societies. This led to her visit to Vienna and Mertens's grandson's recollections. In Erlangen she supervised two PhD students, although officially they were registered under her father. Then Hilbert and Klein invited her to Göttingen, which had become a world-renowned centre for mathematical research. This was 1915, and Hilbert was moving into mathematical physics, inspired by Einstein's theories of relativity. Relativity rests on the mathematics of invariants, though in a more analytic context than the algebraic invariants that Gordan, Hilbert, and Noether had been studying. Namely differential invariants, which include what by then had become basic physical concepts, such as the curvature of space.

Hilbert wanted an expert on invariants, and Noether fitted the bill perfectly. Within a short time she solved two key problems. The first was a method for finding all differential covariants for vector and tensor fields on a Riemannian manifold – in effect, to discover what other quantities behave like Riemann's curvature tensor. This was vital because Einstein's approach to physics was based on the 'relativity' principle that the laws should be the same for any observer, when expressed in any coordinate frame moving at uniform velocity. So the laws ought to be invariant under the transformation group defined by moving frames. The second was an offshoot of this problem. The natural symmetry group for Special Relativity is the Lorentz group, defined by transformations that mix up space and time but preserve the speed of light, giving relativity its unique flavour. Noether proved that every 'infinitesimal transformation' of the Lorentz group gives rise to a corresponding conservation theorem.

We can appreciate Noether's ideas in the more familiar context of Newtonian mechanics, where they also apply and provide significant insights. Classical mechanics boasts several conservation laws, the most familiar being conservation of energy. A mechanical system is any set of bodies that moves, as time passes, according to Newton's laws of motion. In such systems there is a concept of energy, which takes several forms: kinetic energy, related to motion; potential energy, resulting from interaction with a gravitational field; elastic energy, such as that contained in a compressed spring; and so on. The law of conservation of energy states that in the absence of friction, however the system moves, consistent with Newton's laws of motion, its total energy remains constant – is conserved. If there's friction, kinetic energy is converted into another kind of energy, heat, and again the total energy is conserved. Heat is 'really' the kinetic energy of vibrating molecules of matter, but in mathematical physics it's modelled in a different manner from the energy of rigid bodies, rods, and springs, so its interpretation differs from that of the other types mentioned. Other conservation laws of classical mechanics include conservation of momentum (mass times velocity) and angular momentum (a measure of spin whose rather technical definition is irrelevant here).

Thanks to Galois (Chapter 12) and those who followed him, the concept of symmetry had been identified with invariance under groups of transformations: collections of operations that can be performed on some mathematical structure, whose effect is to leave that structure apparently unchanged. An equation has symmetry when some such transformation, applied to a solution of the equation, always yields another solution. The laws of physics, when expressed as mathematical equations, have many symmetries. Newton's laws of motion, for example, have the symmetries of the Euclidean group, which consists of all rigid motions of space. They're also symmetric under time-translation – measuring time from a different starting point – and in some circumstances under time-*reflection*: reversing the direction in which time flows.

Noether's insight was the existence of a link between some types of symmetry and conservation laws. She proved that every *continuous* symmetry – belonging to a family of symmetries corresponding to continuously varying real numbers – gives rise to a conserved quantity.

Let me unpack that, because as stated it's rather enigmatic. Some types of symmetry naturally live in continuous families. The rotations of a plane, for example, correspond to the angle of rotation, which can be any real number. These rotations form a group, whose elements correspond to the real numbers. A technical issue is worth noting: real numbers that differ by a full circle (360° or 2π radians) define the same rotation. All of these 'one-parameter groups' either look like the real numbers, or angles. Translations of space in a given direction, which can be obtained by sliding space rigidly through any distance in the direction concerned, are also continuous symmetries. Other symmetries may be isolated, not part of such a family. Reflection is an example. You can't perform half a reflection, or one tenth of a reflection, so it's not part of any one-parameter group of rigid motions. The infinitesimal transformations that Noether studied in her PhD are another way to think of one-parameter groups. The underlying concept is that of a Lie group and its associated Lie algebra, named after the Norwegian mathematician Sophus Lie.

In Newtonian mechanics, the conserved quantity corresponding to the one-parameter group of time translations turns out to be energy. This reveals a remarkable link between energy and time, which in turns shows up in the uncertainty principle of quantum mechanics. This allows a quantum system to borrow energy (which is temporarily not conserved) provided it pays it back again before nature notices the discrepancy (wait a split second and it *is* conserved). The conserved quantity corresponding to a one-parameter group of spatial translations turns out to be momentum in the corresponding direction, and that for rotations is angular momentum. In short: the fundamental conserved quantities of Newtonian mechanics all come from continuous symmetries of Newton's laws of motion – one-parameter subgroups of the Euclidean group. And the same is true for relativity and, to some extent, for quantum mechanics.

Not bad for a mathematician considered incapable of lecturing in her own right, who had only recently started to work on the problem.

On the strength of this and other successes, Hilbert and Klein battled to convince the university to change its mind about female faculty members. Academic politics as well as inbuilt misogyny came into play, and the professors in the Philosophy Department were vehemently opposed. If a woman could gain Habilitation and charge fees for lectures, what would stop her becoming a professor and a member of the university senate? Heaven forfend! World War I was in full swing, and this gave them a new argument: 'What will our soldiers think when they return to the university and find they are expected to learn at the feet of a woman?'

Hilbert's reply was scathing. 'Gentlemen: I do not see that the sex of the candidate is an argument against her admission as Privatdozent. After all, the Senate is not a bathhouse.' But even that failed to move the philosophers from their entrenched position. Hilbert, inventive and iconoclastic as ever, found a solution. A notice for the Winter semester of 1916–17 reads:

Mathematical Physics Seminar
Professor Hilbert, with the assistance of Dr E. Noether
Mondays from 4 to 6, no tuition.

Noether spent four years lecturing under Hilbert's name, until the university finally caved in. Her Habilitation was approved in 1919, allowing her to obtain the rank of Privatdozent. She remained a leading member of the department until 1933.

We can gauge Noether's lecturing abilities from a trick that her despairing students once played. Usually there were only five to ten students in the class, but one morning she turned up to find a hundred of them. 'You must have the wrong class,' she told them, but no, they insisted, they were there intentionally. So she delivered her lecture to this unusually large gathering.

When she finished, one of her regular students passed her a note. 'The visitors have understood the lecture just as well as any of the regular students.'

The problem with her lectures was straightforward. Unlike most mathematicians, she was a formal thinker. To her, the symbols *were* the concepts. To follow her lectures, you had to think the same way. And that was hard.

Despite this, it was Noether, and her emphasis on formal structures, that would open up much of today's mathematics. Sometimes you just have to bite the bullet.

With Habilitation safely behind her, Noether promptly changed fields, picking up where Dedekind left off when he replaced Kummer's obscure notion of an ideal number by a conceptually simpler but more abstract notion, that of an ideal. The context for this approach was itself abstract: the theory of rings – algebraic systems in which addition, subtraction, and multiplication are defined, and satisfy the usual rules, with the possible exception of the commutative law $xy = yx$ for multiplication. The integers, the real numbers, and polynomials in one or more variables all form rings.

We can get a brief flavour of the set-up using ordinary integers.

The traditional way to think about prime numbers and divisibility is to work with specific integers, such as 2, or 3, or 6. We observe that $6 = 2 \times 3$, so 6 is not prime; on the other hand, no such decomposition into smaller numbers is possible for 2 or 3, so they are prime. But, as Dedekind realised, there's another way to see this. Consider the sets formed by all multiples of 6, 2, and 3, which I'll denote like this:

$$[6] = \{\ldots, -12, -6, 0, 6, 12, 18, 24, \ldots\}$$

$$[2] = \{\ldots, -4, -2, 0, 2, 4, 6, 8, 10, 12, 14, 16, 18, 20, 22, 24, \ldots\}$$

$$[3] = \{\ldots, -6, -3, 0, 3, 6, 9, 12, 15, 18, 21, 24, \ldots\}$$

Here the curly brackets indicate sets, and we allow negative multiples. Observe that *every* member of [6] is a member of [2]. This is obvious: any multiple of 6 is automatically a multiple of 2 because 6 is a multiple of 2. Similarly, every member of [6] is a member of [3]. In other words, you can spot divisors of a given number (here 6) by seeing which sets of this kind contain all multiples of 6.

On the other hand, some numbers in [3] aren't in [2], and conversely. So 2 doesn't divide 3 and 3 doesn't divide 2.

With a bit of fiddling around, the entire theory of primes and divisibility for integers can be reformulated in terms of these sets of multiples of a given number. The sets are examples of ideals, which are defined by two main properties: the sum and difference of numbers in the ideal are also in the ideal, and the product of a number in the ideal by any number in the ring is in the ideal.

Noether restated Hilbert's theorems about invariants in terms of ideals, and then generalised his results in a totally new direction. Hilbert's finite basis theorem for invariants boils down to proving that an associated ideal is finitely generated; that is, it consists of all combinations of a finite number of polynomials (the basis). Noether reinterpreted the argument as the statement that any chain of ever larger ideals must stop after finitely many steps. That is, *every* ideal in the ring of polynomials is finitely generated. She published this idea in 1921 in a far-reaching paper, 'Theory of ideals in ring domains'. This paper kicked off general commutative ring theory.

Noether became adept at squeezing important theorems out of chain conditions, and a ring that satisfies this 'ascending chain condition' is said to be Noetherian. This conceptual approach to invariants was a huge contrast to the turgid calculations in her thesis, which she now dismissed as 'Formelgestrüpp' – a formula jungle.

Today every mathematics undergraduate is taught the abstract axiomatic approach to algebra. Here the most important concept is that of a group, now stripped of all associations with permutations or the solution of algebraic equations. Indeed, an abstract group need not even be composed of transformations. It is defined to be any system of elements that can be combined to yield another element of the system, subject to a short list of simple conditions: the associative law, the existence of an 'identity element' which combines with any other element to yield that element, and the existence for each element of an 'inverse' element, which composes with it to yield the identity. That is, there is an element that has no effect, to each element corresponds another that undoes whatever the element itself does, and if you combine three elements in a row it doesn't matter which pair you combine first.

Slightly more elaborate structures bring into play the full panoply of arithmetical operations. I've already mentioned a ring. There's also a field, in which division is possible too. The precise development of this abstract view is complicated, and many figures contributed to it. Who first did what is often unclear. By the time the precise definitions had been sorted out, most mathematicians already had a pretty clear feeling for what was going on. But at root we owe the whole viewpoint to Noether, who emphasised the need for an axiomatic approach to *all* mathematical structures.

In 1924, Dutch mathematician Bartel van der Waerden joined her circle and became the leading expositor of her approach, encapsulated in his 1931 *Modern Algebra*. By 1932, when she delivered a plenary address at the International Congress of Mathematicians, her algebraic ability was recognised worldwide. She was quiet, modest, and generous. Van der Waerden summed up her contribution when he wrote her obituary:

> The maxim by which Emmy Noether was guided throughout her work might be formulated as follows: Any relationships between

numbers, functions, and operations become transparent, generally applicable, and fully productive only after they have been isolated from their particular objects and been formulated as universally valid concepts.

Noether had more than algebra in her sights. She imported the same insights into topology. To the early topologists, a topological invariant was a combinatorial object, such as the number of independent cycles – closed loops with certain properties. Poincaré had begun the process of adding extra structure, with the concept of homotopy. When Noether found out what the topologists were doing, she immediately spotted something they'd all missed: an underlying abstract algebraic structure. Cycles weren't just things you could count: with a bit of care, you could turn them into a group. Combinatorial topology became algebraic topology. Her viewpoint won instant converts, in particular Heinz Hopf and Pavel Alexandrov. Similar ideas occurred independently to Leopold Vietoris and Walther Mayer in Austria between 1926 and 1928, leading them to define a homology group – a basic invariant of a topological space. Algebra had taken over from combinatorics, revealing a far richer structure that topologists could exploit.

In 1929 Noether visited Moscow State University, to work with Alexandrov and to teach abstract algebra and algebraic geometry. Although not politically active, she quietly expressed support for the Russian Revolution because of the opportunities it opened up in science and mathematics. This didn't go down terribly well with the authorities, who evicted her from her lodgings when students complained about the presence of a Marxist-sympathising Jewess.

In 1933, when the Nazis dismissed Jews from university positions, Noether first tried to obtain a position in Moscow, but eventually moved to Bryn Mawr University in the United States, with the aid of the Rockefeller Foundation. She also gave lectures at the Institute for Advanced Study in Princeton, but complained that even in America she felt uncomfortable in a 'men's university, where nothing female is admitted'.

Despite that, she enjoyed America, but not for long. She died in

1935, from complications after a cancer operation. Albert Einstein wrote in a letter to the *New York Times*:

> In the judgment of the most competent living mathematicians, Fräulein Noether was the most significant creative mathematical genius thus far produced since the higher education of women began. In the realm of algebra, in which the most gifted mathematicians have been busy for centuries, she discovered methods which have proved of enormous importance in the development of the present-day younger generation of mathematicians.

Not only that: she took the men on at their own game, and beat them.

21
The Formula Man
Srinivasa Ramanujan

Srinivasa Ramanujan
Born: Erode, Tamil Nadu, India , 22 December 1887
Died: Kumbakonam, Tamil Nadu, India, 26 April 1920

IT WAS JANUARY 1913. Turkey was at war in the Balkans and Europe was being dragged deeper and deeper into the conflict. Godfrey Harold Hardy, professor of mathematics at Cambridge University, despised war; and he took great pride that the field of his life's work, pure mathematics, had no military uses.

Outside, snow drizzled damply down, while begowned undergraduates scurried through the slush of Trinity Great Court. But in Hardy's rooms a cheerful fire kept the cold at bay. On the table lay the morning post, ready to be opened. He glanced at the envelopes. One caught his eye because of its unusual postage stamps. India. Postmarked Madras, 16 January 1913. Hardy slit the manilla envelope, more than a little battered by its long journey, and drew out a sheaf of papers. An accompanying letter, in an unfamiliar hand, began:

> Dear Sir,
>   I beg to introduce myself to you as a clerk in the Accounts Department of the Port Trust Office at Madras at a salary of only £20 per annum. I am now about 23 years of age. I have had no University education ... After leaving school I have been employing the spare time at my disposal to work at Mathematics ... I am striking out a new path for myself.

Oh Lord, another crank. Probably thinks he's squared the circle. Hardy nearly threw the letter in the wastebasket, but as he picked it up a sheet of mathematical symbols caught his eye. Curious formulas. A few, he recognised. Others were ... unusual.

If the author of the letter is a crank, he might at least prove an entertaining crank. Hardy read on:

> Very recently I came across a tract by you styled *Orders of Infinity* in page 36 of which I find a statement that no definite expression has as yet been found for the number of prime numbers less than any given number. I have found an expression which very nearly approximates to the real result, the error being negligible.

My word. He's rediscovered the Prime Number Theorem.

> I would request you to go through the enclosed papers. Being poor, if you are convinced there is anything of value I would like to have my theorems published ... Being inexperienced I would very highly value any advice you give me. Requesting to be excused for the trouble I give you.
>   I remain, Dear Sir, Yours Truly,
>   S. Ramanujan

Not a typical crank, Hardy mused. *A typical crank would be more aggressive and more conceited.* Putting the letter aside, he picked up the enclosed sheets and began to read. Half an hour later he was sitting back in his chair with an odd expression on his face. *How strange.* Hardy was intrigued. But it was time for his undergraduate lecture on analysis, so he shrugged into his chalk-spattered gown, walked out of the room, and shut the door behind him.

That evening, over high table, he talked about the strange letter to any of the Fellows of the college who cared to listen, including his colleague and close collaborator John Littlewood. Littlewood was willing to waste an hour on the matter to help put his friend's mind at ease, and the chess room was free. As they walked in, Hardy held up the slim sheaf of paper. 'This man,' he announced to the gathering at large, 'is either a crank or a genius.'

An hour later, Hardy and Littlewood emerged with the verdict. *Genius.*

I hope you'll forgive my dramatisation of these events. I've put thoughts into Hardy's head, but surviving documentation makes it clear that something very similar must have gone through his mind, and the general turn of events respects recorded history.

The author of the letter, Srinivasa Ramanujan, was born into a Brahmin family in 1887. His father K. Srinivasa Iyengar was a clerk in a sari shop, and his mother Komalatammal was a bailiff's daughter. The birth took place in his grandmother's house in Erode, a town in the southern province of Tamil Nadu, India. He grew up in Kumbakonam, where his father worked. But it was common for a young wife to spend time with her parents as well as her husband, so his mother frequently took him to live with her father near Madras, some 400 kilometres away. The family was poor, the house tiny. It was basically a happy childhood, although Ramanujan was very obstinate. For the first three years of his life, he scarcely said a word, and his mother feared he was dumb. Aged five, he didn't like his teacher and didn't want to go to school. He preferred to think about things for himself, asking annoying questions such as 'How far apart are clouds?'

Ramanujan's mathematical talents surfaced early, and by the age of 11 he had outstripped two college students who lodged at his home. He learned how to solve cubic equations and could recite the digits of π and e at some length. A year later he borrowed an advanced textbook and mastered it completely, without apparent effort. When he was 13 he devoured Sidney Loney's *Trigonometry*, which included the infinite series expansions for the sine and cosine, and was already producing his own new results. His ability in mathematics won him many prizes at school, and in 1904 the headmaster described him as deserving more marks than the maximum possible.

At the age of 15 an event occurred which was to change his life, but at the time it seemed mundane. He borrowed a copy of George Carr's *Synopsis of Elementary Results in Pure Mathematics* from the Government College Library. The *Synopsis* is, to say the least, idiosyncratic. Its thousand-plus pages list some five thousand theorems – all without proofs. Carr based the book on problems that he posed when coaching his students. Ramanujan likewise set himself a problem: to establish *all the formulas in the book*. He had no help, no other books. Effectively, he'd set himself a research project of five thousand separate topics. Too poor to afford paper, he did his calculations on a slate and jotted the results in a series of notebooks, which he kept throughout his life.

In 1908 Ramanujan's mother Komalatammal decided to find her son, then aged twenty, a wife. She settled on Janaki, the daughter of one of her relatives, who lived about 100 kilometres from Kumbakonam. Janaki was nine. The age difference wasn't a great obstacle in a society of arranged marriages and child brides. Ramanujan was – it seemed – a very ordinary young man; a lazy failure with no job, no money, and no prospects. But Janaki was one of five daughters in a family that had lost most of what it owned, and her parents would be happy merely to find a husband who would be kind to her. That was enough for Komalatammal, which ordinarily meant it was a done deal. But on this occasion her husband blew his top. His son could do better! He had nearly married two years before, but by mischance a death in the bride's family put paid to it. Mostly, the father was upset that his wife hadn't asked his advice first. At any rate, he snubbed the bride's family by refusing to come to the wedding.

The wedding day dawned, and there was no sign of the groom or his family. The bride's father Rangaswamy announced to all and sundry that if Ramanujan didn't turn up soon, he'd marry Janaki off on the spot to someone else. Eventually the train from Kumbakonam turned up, hours late, and it was well after midnight when Ramanjuan and his mother (minus father) arrived at the village on a bullock cart. Komalatammal quickly made short work of Rangaswamy's threats, pointing out very publicly that a poor father with five daughters would reject a genuine offer at his peril.

After the customary five to six days of celebrations, Janaki found herself married to Ramanujan. She wouldn't join him until she reached puberty, but both their lives had changed. Ramanujan started looking for a job. He tried tutoring students in mathematics, but found no takers. When he became ill, possibly as a result of a previous operation, he turned up in a horse cart at the home of a friend, R. Radhakrishna Iyer, who took him to see a doctor and then put him on a train back to Kumbakonam. Just as he was leaving, Ramanujan said 'If I die, please hand these over to Professor Singaravelu Mudaliar or to the British professor Edward Ross.' And he pressed into his startled friend's hands two fat notebooks, stuffed full of mathematics.

Here was not just Ramanujan's legacy, but his job ticket: evidence that he was more than an indolent wastrel. He started calling on influential people with his mathematical portfolio under his arm. In *The Man Who Knew Infinity*, Robert Kanigel says: 'Ramanujan had become, in the year and a half since his marriage, a door-to-door salesman. His product was himself.' It was a hard sell. In India at that time, the best route to employment was the right connections, but Ramanujan had none. All he had was his notebooks ... and one other important thing. He was friendly. Everyone liked him. He was lively, and told jokes.

Eventually, his persistence and uncomplicated charm paid off. In 1912 a mathematics professor, P.V. Seshu Aiyar, sent him to see R. Ramachandra Rao, a civil servant who was a district collector at Nellore. Rao recollected the interview:

> I condescended to permit Ramanujan to enter my presence. A short uncouth figure, stout, unshaved, not overclean, with one

conspicuous feature – shining eyes ... I saw quite at once that there was something out of the way; but my knowledge did not permit me to judge whether he talked sense or nonsense ... He shewed me some of his simpler results. These transcended existing books and I had no doubt that he was a remarkable man. Then, step by step, he led me to elliptic integrals and hypergeometric series and at last his theory of divergent series not yet announced to the world converted me.

Rao secured Ramanujan an appointment in the Madras Port Trust Office at 30 rupees per month, a job that left him enough spare time to continue his researches. Another bonus was that he could take away used wrapping paper to write his mathematics on.

It was then that, at the urgings of the same people, Ramanujan wrote his diffident letter to Hardy. Hardy immediately sent an encouraging reply. Ramanujan asked him to send a 'sympathetic letter' to help him get a scholarship. Hardy was ahead of him, and more ambitious. He'd already written to the Secretary for Indian Students in London, seeking a way to get Ramanujan a Cambridge education. But then it transpired that Ramanujan didn't want to leave India. The Cambridge network rolled into action. Another Trinity mathematician, Gilbert Walker, was visiting Madras, and he wrote a letter to the University of Madras, which granted Ramanujan a special scholarship. At last he was free to devote all of his time to mathematics.

Hardy continued trying to persuade Ramanujan to come to England. Ramanujan began to waver, and the main obstacle became his mother. Then, one morning, to general family astonishment, she announced that the goddess Namagiri had appeared to her in a dream, commanding her to let her son fulfil his life's calling. Ramanujan was given a grant to cover subsistence and travel, set sail for England, and by April 1914 was in Trinity College. He must have felt very out of place, but he stuck to it, publishing many research papers, including some influential joint work with Hardy.

Ramanujan was Brahmin, a Hindu caste forbidden to cause harm to living creatures. Although his English friends got the impression that his main religious motivation was not belief, but social custom, he observed the proper rituals as far as was possible in wartime England. As a vegetarian, he didn't trust the College cooks to eliminate all meat products, so he taught himself to cook, Indian style of course. According to friends, he became an excellent cook.

Around 1916 his friend Gyanesh Chandra Chatterji, in Cambridge as a Government of India state scholar, was about to get married, so Ramanujan invited him and his bride-to-be for dinner. As agreed, Chatterji, fiancée, and a chaperone turned up at Ramanujan's rooms, and he served them soup. When they polished it off, he offered some more, and all three took a second helping. So he suggested a third helping. Chatterji accepted, but the ladies declined.

Shortly afterwards, Ramanujan was nowhere to be seen.

They waited for him to return. After an hour had passed, Chatterji went downstairs to find a porter. Yes, he had seen Mr Ramanujan. He had called a taxi, and gone off in it. Chatterji returned to the room and the three guests waited until ten o'clock at night, when college rules required them to leave. No sign of their host. No sign of him for the next four days ... What had happened? Chatterji was worried.

On day five, a telegram arrived from Oxford: could Chatterji wire Ramanujan five pounds? (That was a lot in those days, a few hundred pounds today.) Money sent, Chatterji waited, and Ramanujan duly appeared. Asked what had happened, he explained. 'I felt hurt and insulted when the ladies didn't take the food I served.'

It was an outward sign of inner turmoil. Ramanujan was at the end of his tether. He had never truly adapted to life in England. His health, never good, was getting worse, and he ended up in hospital. Hardy visited him there, and the visit led to another story about Ramanujan that also features a taxi. It has become something of a cliché, but it bears repetition all the same.

Hardy once wrote that every positive integer was one of Ramanujan's personal friends, and illustrated this with an anecdote about visiting Ramanujan in hospital. 'I had ridden in taxi cab number 1729 and remarked that the number seemed to me rather a dull one, and that I hoped it was not an unfavourable omen. "No," he replied, "it is a very interesting number; it is the smallest number expressible as the sum of two cubes in two different ways."'

To be precise,

$$1729 = 1^3 + 12^3 = 9^3 + 10^3$$

and it is the smallest positive number with such a property.

The story makes its point well, but I can't help wondering whether it was a bit of a set-up, with Hardy trying to buck up his sick friend by getting him to rise to the bait. Most people wouldn't spot this feature of the number 1729, to be sure, but Ramanujan would undoubtedly recognise it immediately. Indeed, many mathematicians, especially those with an interest in number theory – Hardy among them – would be aware of it. It's almost impossible for a mathematician to look at 1729 and not think of 1728, which is the cube of 12. And it's also hard not to notice that 1000 is 10 cubed and 729 is 9 cubed.

Be that as it may, Hardy's story led to a minor but intriguing concept in number theory: that of a taxicab number. The nth taxicab number is the smallest number that can be expressed as a sum of two positive cubes in n distinct ways. The next two taxicab numbers are

$$87,539,319$$

$$6,963,472,309,248$$

There are infinitely many taxicab numbers, but only the first six are known.

By 1917 Ramanujan was back in his rooms, obsessed with mathematics to the exclusion of all else. He would work day and night, then collapse exhausted and sleep for 20 hours. This did his health no good, and the war caused shortages of the fruit and vegetables on which he relied. By spring, he was afflicted by some undiagnosed but probably incurable disease. He was admitted to a small private hospital for patients from Trinity College. Over the following two years he saw eight or more doctors and was admitted to at least five hospitals and sanatoriums. The doctors suspected a gastric ulcer, then cancer, then blood poisoning; but they decided that the most likely cause was tuberculosis, and this is what they mainly treated him for.

Finally, much too late, academic honours were coming his way. He became the first Indian to be elected a Fellow of the Royal Society, and Trinity elected him to a Fellowship. Reinvigorated, he picked up his mathematics again. But his heath remained poor, the English climate was suspect, and in April 1919 he returned to India. The long voyage didn't suit him, and by the time he arrived in Madras his health had once more deteriorated. In 1920 he died in Madras, leaving a widow but no children.

There are four main sources for Ramanujan's mathematics: his published papers, his three bound notebooks, his quarterly reports to the University of Madras, and his unpublished manuscripts. A fourth 'lost' notebook – a bundle of loose sheets – was found again in 1976 by George Andrews, but some of his manuscripts are still missing. Bruce Berndt has edited a three-volume work *Ramanujan's Notebooks*, including proofs of all of his formulas.

Ramanujan had an unusual background, and no formal training. It was hardly a suprise that his mathematics was a little idiosyncratic. His greatest strength was in an unfashionable area – the production of ingenious and intricate formulas. Ramanujan was the Formula Man *par excellence*, unrivalled by any save a few Old Masters such as Euler and Jacobi. 'There is always more in one of Ramanujan's formulae than meets the eye,' Hardy wrote. Most of his results are about infinite series, integrals, and continued fractions. An example of a continued fraction is the expression:

$$\cfrac{x}{1+\cfrac{x^5}{1+\cfrac{x^{10}}{1+\cfrac{x^{15}}{1+\ldots}}}}$$

which was on the last page of his letter, featuring in a distinctly weird, but correct, formula. He applied some of his formulas to number theory, taking a special interest in analytic number theory, which seeks simple approximations to such quantities as the number of primes below a given limit – Gauss's prime number theorem (Chapter 10) – or the average number of divisors of a given number.

His publications while at Cambridge were influenced by his contact with Hardy, and written in a conventional style with rigorous proofs. The results recorded in his notebooks have a very different quality. Because he was self-educated, his concept of proof was less than rigorous. If a mixture of numerical evidence and formal argument led to a plausible conclusion, and his intuition told him he had the right answer, that was enough for Ramanujan. His results were usually correct, but his proofs often had gaps. Sometimes any competent technician could fill the gaps, and sometimes quite different arguments were needed. On rare occasions, his results were wrong. Berndt argues that if Ramanujan 'had thought like a well-trained mathematician, he would not have recorded many of the formulas which he thought he had proved', and mathematics would have been poorer as a result.

A good example is a result that Ramanujan called his 'Master Formula'.[11] His proof involves series expansions, interchanges of the order of summation and integration, and other similar manoeuvres. Because he uses infinite processes, each step is fraught with danger. The greatest analysts spent most of the nineteenth century working out just when such procedures are permissible. The conditions that, according to Ramanujan, make his formula true, are grossly insufficient. Nevertheless, almost all of the results that he derives from his Master Formula are correct.

Some of Ramanujan's most striking work is in the theory of partitions, a branch of number theory. Given a whole number, we ask in how many ways it can be partitioned, that is, written as a sum of smaller whole numbers. For example, the number 5 can split up in seven ways:

$$5 \quad 4+1 \quad 3+2 \quad 3+1+1 \quad 2+2+1$$
$$2+1+1+1 \quad 1+1+1+1+1$$

Therefore $p(5) = 7$. The numbers $p(n)$ grow rapidly with n. For instance, $p(50) = 204,226$ and $p(200)$ is a staggering 3,972,999,029,388. No simple formula for $p(n)$ exists. However, we can ask for an approximate formula, giving the general order of magnitude of $p(n)$. This is a problem in analytic number theory, and an especially intractable one. In 1918 Hardy and Ramanujan overcame the technical difficulties and derived an approximate formula, a rather complicated series involving complex 24th roots of unity. They then found that when $n = 200$ the first term *alone* agrees with the first six significant figures of the exact value. By adding just seven more terms they obtained 3,972,999,029,388·004, whose integer part is the exact value. They observed that this result 'suggests very forcibly that it is possible to obtain a formula for $p(n)$ which not only exhibits its order of magnitude and structure, but may be used to calculate its exact value for any n', and they went on to prove precisely that. It must be one of the very few occasions when the search for an approximate formula led to an exact one.

Ramanujan also found some remarkable patterns in partitions. In 1919 he proved that $p(5k + 4)$ is always divisible by 5 and $p(7k + 5)$ is always divisible by 7. In 1920 he stated some similar results: for example $p(11k + 6)$ is always divisible by 11; $p(25k + 24)$ is divisible by 25; all of $p(49k + 19)$, $p(49k + 33)$, $p(49k + 40)$, and $p(49k + 47)$ are divisible by 49; and $p(121k + 116)$ is divisible by 121. Notice that $25 = 5^2$, $49 = 7^2$, and $121 = 11^2$. Ramanujan said that, as far as he could tell, such formulas exist only for divisors of the form $5^a 7^b 11^c$, but this was wrong. Arthur Atkin found that $p(17303 + 237)$ is divisible by 13, and in 2000 Ken Ono proved that congruences of this kind exist for all prime moduli. A year later he and Scott Ahlgren proved they exist for all moduli not divisible by 6.

Some of Ramanujan's results remain unproved even to this day. One that succumbed about forty years ago is particularly significant. In a paper of 1916 he studied a function $\tau(n)$ defined to be the coefficient of x^{n-1} in the expansion of

$$\left[(1-x)\left(1-x^2\right)\left(1-x^3\right)...\right]^{24}$$

Thus $\tau(1) = 1$, $\tau(2) = -24$, $\tau(3) = 252$, and so on. The formula comes from deep and beautiful work in the nineteenth century on elliptic functions. Ramanujan needed $\tau(n)$ to solve a problem about powers of divisors of n, and he needed to know how big it was. He proved that its size is no larger than n^7, but conjectured that this can be improved to $n^{11/2}$. He conjectured two formulas:

$\tau(mn) = \tau(m)\tau(n)$ if m and n have no common factor
$\tau\left(p^{n+1}\right) = \tau(p)\tau(p^n) - p^{11}\tau\left(p^{n-1}\right)$ for all prime p

These make it easy to compute $\tau(n)$ for any n. Louis Mordell proved them in 1919, but Ramanujan's conjecture on the order of magnitude of $\tau(n)$ resisted all his efforts.

In 1947 André Weil was looking over old results of Gauss, and he realised he could apply them to integer solutions of various equations. Following his nose, and a curious analogy with topology, he formulated a series of rather technical results, the Weil conjectures. These acquired a central position in algebraic geometry. In 1974 Pierre Deligne proved them, and a year later he and Yasutaka Ihara deduced Ramanujan's conjecture from them. That his innocent-looking conjecture required such a massive and central breakthrough before it could be answered is a sign of how good Ramanujan's intuition was.

Among his more enigmatic inventions were 'mock theta functions', which he described in his final letter to Hardy in 1920; details were later found in his lost notebook. Jacobi introduced theta functions as an alternative approach to elliptic functions. They're infinite series that transform in a very simple way when suitable constants are added to the variable, and elliptic functions can be constructed by dividing one theta function by another. Ramanujan defined some analogous series, and stated a large number of formulas involving them. At the time, the whole idea seemed to be just an exercise in manipulating complicated series, with no connection to anything else in mathematics. Today, we realise this is not the case. They have important connections with the theory of modular forms, which arise in number theory and are also related to elliptic functions.

A similar but distinct concept, the Ramanujan theta function, has recently turned out to be useful in string theory, the most popular attempt by physicists to unify relativity and quantum mechanics.

Because Ramanujan functioned in such an extraordinary manner, obtaining correct results by non-rigorous methods, it has sometimes been suggested that his thought patterns were special or unusual. Ramanujan himself is quoted as saying that the goddess Namagiri told them to him in dreams. However, he may have said this just to avoid embarrassing discussions. According to his widow S. Janaki Ammal Ramanujan, he 'never had time to go to the temple because he was constantly obsessed with mathematics'. Hardy wrote that he believed 'all mathematicians think, at bottom, in the same way, and Ramanujan was no exception', but he added: 'He combined a power of generalisation, a feeling for form, and a capacity for rapid modification of his hypotheses, that were often really startling.'

Ramanujan wasn't the greatest mathematician of his period, nor the most prolific; but his reputation does not just rest on his remarkable background and touching 'poor boy makes good' story. His ideas were influential during his lifetime, and they grow more influential as the years pass. Bruce Berndt believes that, far from being old-fashioned, Ramanujan was ahead of his time. It's sometimes easier to prove one of Ramanujan's remarkable formulas than it is to work out how he could possibly have thought of it. And many of Ramanujan's deepest ideas are only now becoming appreciated. I leave the final word to Hardy:

> One gift [that his mathematics] has which no one can deny:
> profound and invincible originality. He would probably have been a
> greater mathematician if he had been caught and tamed a little in
> his youth; he would have discovered more that was new, and that,
> no doubt, of greater importance. On the other hand he would have
> been less of a Ramanujan, and more of a European professor, and
> the loss might have been greater than the gain.

22

Incomplete and Undecidable
Kurt Gödel

Kurt Friedrich Gödel
Born: Brünn, Austria-Hungary, 28 April 1906
Died: Princeton NJ, USA, 14 January 1978

THE STEREOTYPIC IMAGE of mathematicians, aside from all of them being male and elderly, is that they're a bit strange. Other-worldly, certainly. Eccentric, commonly. Downright crazy, sometimes.

We've seen that this image doesn't fit most mathematicians, aside from their being male, and even that has changed dramatically over the last few decades. Agreed, mathematicians tend to end their careers by being elderly, but who doesn't? The only way to avoid this is to die young, like Galois. Reputations and responsibilities tend to grow with age, so the elderly are likely to be over-represented among the leaders of the subject.

When their minds are focused on research, mathematicians can easily appear other-worldly, but as a biologist colleague of mine always insisted, they're not absent-minded: they're present-minded somewhere else. If you want to solve a difficult mathematical problem, you need to concentrate. Some mathematicians (by no means the only profession to do so) take this lack of worldly focus to the point of eccentricity. Perhaps the most obvious example is Paul Erdős, who never held an academic position and never owned a house. He travelled from one colleague to another, spending a night on the sofa or months in the spare room. Yet he wrote an extraordinary 1500 research papers and collaborated with a staggering 500 different mathematicians.

As for being crazy: some are, at some stage of their life, mentally ill. Cantor suffered from serious bouts of depression. John Nash, the subject of the book and movie *A Beautiful Mind*, won the 1994 Nobel Prize in Economics (more precisely, the Nobel Memorial Prize, which is treated like any of the original Nobel Prizes for most purposes). Yet he suffered for many years from a condition diagnosed as paranoid schizophrenia, and underwent electroshock therapy. By an effort of will, recognising psychotic interludes and refusing to give in to them, he managed to cure himself.

Kurt Gödel was definitely eccentric, and at times went beyond. His chosen area of mathematical logic was not, at that time, mainstream mathematics, and in this respect he was if anything more other-worldly than most of his colleagues. In compensation, his discoveries in that area revolutionised our thinking about the foundations of logic and mathematics, and how these interact. He was brilliantly original, and remarkably deep.

His interest in logic began in 1933 when Adolf Hitler rose to power in Germany, and was stimulated by seminars given by Moritz Schlick, a philosopher who founded logical positivism and the Vienna Circle. In 1936 one of Schlick's former students, Johann Nelböck, murdered him. Many members of the Vienna Circle had already fled Germany, fearing anti-Semitic persecution, but Schlick, who was in Austria, stayed on at the University of Vienna. He was walking up the steps to give a lecture when Nelböck shot him with a pistol. Nelböck confessed to the murder, but used the court proceedings as a platform from which to proclaim his political beliefs. He claimed his lack of moral restraint had been a reaction to Schlick's philosophical stance, which was antagonistic to metaphysics. Others suspected the true cause to have been Nelböck's infatuation with another student, Sylvia Borowicka. His unrequited passion led him to a paranoid belief that Schlick was a competitor for her affections. He was sentenced to ten years in prison, but the case contributed to a growing anti-Semitic hysteria in Vienna, even though Schlick was not, in fact, Jewish. Post-truth politics is nothing new. Worse, when Germany annexed Austria, Nelböck was released, a mere two years into his sentence.

The murder of his mentor had a terrible effect on Gödel. He, too, developed signs of paranoia – though it rather fitted the old joke 'Just because I'm paranoid, it doesn't mean they're not out to get me.' Gödel wasn't Jewish either, but he had many friends who were. Living under Nazi rule, paranoia was the ultimate in sanity. However, he developed a phobia about being poisoned, and spent several months being treated for mental illness. This fear came back to haunt him in the last few years of his life, when he again developed symptoms of mental illness and paranoia. He refused to eat any food that his wife hadn't cooked. In 1977 she had two strokes, and was admitted to hospital for a lengthy period, so she was no longer able to cook for him. He stopped eating, and starved himself to death. It was a gruesome and futile end for one of the greatest thinkers of the twentieth century.

Gödel's father Rudolf managed a textile factory in Brünn, Austria-Hungary, now Brno in the Czech Republic. From early childhood and well into adulthood he was very close to his mother Marianne (*née* Handschuh). Rudolf was Protestant, Marianne Catholic; Kurt was brought up in the Protestant church. He considered himself a committed Christian, believing in a personal God, but not in organised religion. He wrote that 'religions are, for the most part, bad – but religion is not'. He read the Bible regularly but didn't attend church. An attempt at a mathematical proof of the existence of God, derived using modal logic, was found among his unpublished papers. His nickname in the family, as a child, was *Herr Warum* (Mr Why), for reasons you can guess. When six or seven years old he suffered a bout of rheumatic fever, and although he made a complete recovery, he never lost the belief that the illness had damaged his heart. His health was often fragile, a state of affairs that continued until his death.

From 1916 Gödel was a student at the Deutsches Staats-

Realgymnasium, gaining high marks in all of his subjects, especially mathematics, languages, and religion. He was automatically made a Czechoslovak citizen when the Austro-Hungarian Empire broke up at the end of World War I. He attended the University of Vienna in 1923, unsure initially whether to study mathematics or physics, but Bertrand Russell's *Introduction to Mathematical Philosophy* led him to settle on mathematics, with the main focus being on mathematical logic. A key turning point in his career happened in 1928, when he went to a lecture in Bologna by David Hilbert at the first International Congress of Mathematicians to be held after the end of World War I. Hilbert explained his views about axiomatic systems, especially whether they're consistent and complete. In 1928 Gödel read *Principles of Mathematical Logic* by Hilbert and Wilhelm Ackermann, which provided the technical backbone for Hilbert's programme to settle these questions. In 1929 he chose that topic for his doctoral thesis, working under Hans Hahn. He proved what we now call Gödel's Completeness Theorem: that predicate calculus (Chapter 14) is complete. That is, every true theorem can be proved, every false one can be disproved, and there's no other option left. However, predicate calculus is very limited, and inadequate as a foundation for mathematics. Hilbert's programme was formulated within a much richer axiomatic system.

Gödel became an Austrian citizen the same year. (His citizenship automatically changed to German in 1938 when Germany annexed Austria.) He was awarded a doctorate in 1930. In 1931 he demolished Hilbert's programme by publishing 'On formally undecidable propositions of *Principia Mathematica* and similar systems', which proved that no axiom system rich enough to formalise mathematics can be logically complete, and that it's impossible to prove any such system is consistent. (I'll tell you about *Principia Mathematica* in a moment.) He attained Habilitation in 1932, becoming a Privatdozent at the University of Vienna in 1933. The harrowing events recounted earlier happened during this period of his life. To get a break from Nazi Austria, he visited the United States. There he met, and became friends with, Einstein.

In 1938 he married Adele Nimbursky (*née* Porkert), whom he had met at the Der Nachtfalter night club in Vienna eleven years earlier. She was six years older than he was, had been married previously, and both his parents objected, but he ignored their wishes. When World War II started in 1939, Gödel became concerned that he might be drafted into the German army. His poor health should have ruled that out, but he'd already been mistaken for being Jewish, so he might also be mistaken for being healthy. He managed to wangle an American visa, and headed for the USA by way of Russia and Japan, along with his wife. They arrived there in 1940. In that year he proved that Cantor's Continuum Hypothesis is consistent with the usual set-theoretic axioms for mathematics. He took up a position at the Institute for Advanced Study in Princeton, first as an ordinary member, then a permanent one, then, from 1953, a professor. Although he stopped publishing in 1946, he continued to do research.

Gödel became a US citizen in 1948. Apparently he believed he'd found a logical flaw in the US Constitution, and attempted to explain it to the judge, who sensibly failed to take the bait. His close friendship with Einstein led him to do some work on relativity. In particular, he found a spacetime that possesses a closed timelike curve – a mathematical formulation of a time machine. If something follows such a curve through space and time, its future merges into its past. It's like being in London in 1900, travelling twenty years into the future, and finding you're back in London and the year is again 1900. More recently, closed timelike curves have become a hot topic, not so much because they might lead to a practical time machine, but because they shed light on the limitations of General Relativity, and suggest a possible need for new laws of physics.

In his final years, Gödel's health, never good, became worse. His brother Rudolf reported that he

> had a very individual and fixed opinion about everything ...
> Unfortunately he believed all his life that he was always right not
> only in mathematics but also in medicine, so he was a very difficult
> patient for doctors. After severe bleeding from a duodenal ulcer ...
> he kept to an extremely strict (over strict?) diet, which caused him
> slowly to lose weight.

What happened after that, you already know. On his death certificate the cause of death is stated as 'malnutrition and inanition caused by personality disturbance'. Inanition is exhaustion caused by lack of food. He weighed just thirty kilograms.

＊

Since ancient times, mathematics was held up as a shining example of something that was simply *true* – absolute truth, no ifs or buts. Two plus two is four: take what you get and no whining. Its sole competitor for absolute truth was religion (denomination and sect of the believer's choice, of course), but mathematics had a sneaky advantage even then. Religions, as Terry Pratchett has said, are true 'for a given value of true'. Mathematics could *prove* it was true.

As philosophers, logicians, and mathematicians inclined in those directions, started to think more deeply about what this type of absolute truth involves, they realised that it's to some extent illusory. Two plus two equals four for whole numbers, but what, exactly, is a number? For that matter, what are 'plus' and 'equals'? Mathematicians answered these questions by formulating the continuum of real numbers, but Kronecker considered this 'the work of man', believing that only the integers are God-given. It's hard to see how an arbitrary creation of the human mind can constitute absolute truth. It has to be a convention, at best.

The notion that mathematics consists of necessary truths was abandoned in favour of them being deductions from explicit assumptions according to some specified system of logic. Honesty then demands following Euclid's lead and stating those assumptions and logical rules as a system of explicit axioms. This is metamathematics – applying mathematical principles to the internal logical structure of mathematics itself. Bertrand Russell and Alfred North Whitehead paved the way in their 1910–13 *Principia Mathematica* – the title was a conscious homage to Newton – and after several hundred pages they managed to define the number 'one'. After that, the pace hotted up, and more advanced mathematical concepts appeared with ever-increasing rapidity until it was obvious that the rest could be done in the same way and they gave up. One technical feature, their theory of 'types', introduced to avoid certain paradoxes, was later abandoned in favour of other axiom schemes for set theory, the most popular being that of Ernst Zermelo and Abraham Fraenkel.

It's against this background that Hilbert sought to complete the logical circle by proving that some such axiomatic system is logically consistent (no proof leads to a contradiction) and complete (every meaningful statement either has a proof or has a disproof). The first step is essential because, in an inconsistent system, 'two plus two equals five' has a proof. Indeed, *any* statement can be proved. The second identifies 'true' with 'has a proof' and 'false' with 'has no proof'. Hilbert focused on an axiomatic system for arithmetic because *Principia Mathematica* deduced everything else in mathematics from that. To pick up on Kronecker, once God has given us the integers, Man can sort out the rest. Hilbert's programme outlined a series of steps that he believed would achieve this goal, based on the logical complexity of the statements concerned, and he managed to sort out some of the simpler cases. It all looked promising.

Gödel, I suspect, spotted something philosophically fishy about the whole enterprise. In effect, an axiomatic system for mathematical logic was being asked to demonstrate its own consistency. 'Are you consistent?' 'Of course I am!' Pause. 'Yeah, yeah ... Why should I believe *you*?' Be that as it may, some source of scepticism led him to prove two devastating results: his incompleteness theorem and his consistency theorem.

The second rests on the first. Bearing in mind that an inconsistent logical system can prove anything, it can presumably prove the statement 'this system is consistent'. (It can of course also prove 'this system is inconsistent', but ignore that.) So what kind of guarantee of truth can such a proof offer? None. That's what the 'yeah, yeah' response intuitively grasps. There's one possible way that Hilbert's programme can escape from this trap: perhaps the statement 'this system is consistent' makes no sense within the formal axiomatic system. Certainly the statement doesn't look much like arithmetic.

Gödel's answer was to *turn it into* arithmetic. A formal mathematical system is built from symbols, and a proof (or alleged proof) of some statement is merely a string of symbols. The symbols can be given code numbers, and a string of them can also be given a unique numerical code. Gödel numbering achieves this by turning a string of code numbers *abcdef...* into a single number defined by multiplying powers of primes:

$$2^a 3^b 5^c 7^d 11^e 13^f \ldots$$

To decode it back into a string, appeal to uniqueness of prime factorisation.

There are other ways to encode symbol strings as numbers: this one is mathematically elegant and totally impractical. But all Gödel needed was its existence.

Not only do statements encode as numbers: so do proofs, which are just sequences of statements. The logical rules for deducing each statement from the previous ones provide constraints on which of these numbers can correspond to a logically valid proof. So the statement 'P is a valid proof of statement S' can itself be thought of as a statement in arithmetic: 'if you decode P into a sequence of numbers, the final one is the number corresponding to S'. Gödel numbering lets us pass from a metamathematical statement about the existence of a proof to an arithmetical statement about the corresponding numbers.

Gödel wanted to play this game with the statement 'this statement is false'. He couldn't do it directly, because that statement isn't arithmetical. But it can be *made* arithmetical using Gödel numbers, and then it effectively becomes 'this theorem has no proof'. There are some technical tricks to make all this sensible, but that's the gist of it. Suppose, now, that Hilbert is right, and the axiomatic system for arithmetic is complete. Then 'this theorem has no proof' either has a proof, or it doesn't. Either way, we're in trouble. If it has a proof, that's a contradiction. If it has no proof, it's false (we're assuming Hilbert is right, remember?), so it does have a proof – another contradiction. So the statement is self-contradictory ... and there's a theorem in arithmetic that can neither be proved nor disproved.

Gödel quickly parlayed this result into his consistency theorem: if an axiomatic formulation of arithmetic is consistent, then no proof of its consistency can exist. This is the 'yeah, yeah' point in its formal glory: if anyone ever found a proof that arithmetic is consistent, we can immediately deduce that it's not.

For a while Hilbert and his followers hoped that Gödel's theorems just indicated a technical deficiency of the particular axiomatic system set up in *Principia Mathematica*. Perhaps some alternative could avoid the trap. But it soon became apparent that the same argument works in *any* axiomatic system rich enough to formalise arithmetic. Arithmetic is inherently incomplete. And if it's logically consistent, which most mathematicians believe and all of us assume as a working hypothesis, you can never prove it is. In a stroke, Gödel changed humanity's entire philosophical view of mathematics. Its truths can't be absolute – because there are statements whose truth or falsity lies outside the logical system altogether.

We generally assume that an unsolved conjecture, like the Riemann Hypothesis, is either true or false, so either there's a proof or there's a disproof. Post-Gödel, we must add a third possibility. Maybe no logical path leads from the axioms of set theory to the Riemann Hypothesis, *and* no logical path leads from the axioms of set theory to the negation of the Riemann Hypothesis. If so, there's no proof that it's true, *and* no proof that it's false. Most mathematicians would bet that the Riemann Hypothesis is decidable. In fact, most think it's true, and that one day a proof will be found. But if not, surely a counterexample will be found instead, a zero off the critical line. The point is, *we don't know that*. We assume that 'sensible' theorems either have proofs or disproofs, while undecidable ones look a bit contrived and artificial. However, in the next chapter we'll see that a sensible natural question in theoretical computer science turned out to be undecidable.

Classical logic, with its sharp distinction between truth and falsity, with no middle ground, is two-valued. Gödel's discovery suggests that for mathematics, a three-valued logic would be more appropriate: true, false, or undecidable.

23

The Machine Stops
Alan Turing

Alan Mathison Turing
Born: London, 23 June 1912
Died: Wilmslow, Cheshire, 7 June 1954

ACCORDING TO HIS COLLEAGUE Jack Good at Bletchley Park, Alan Turing suffered from hay fever. He cycled into the office, and every June he wore a gas mask to protect him from the pollen. There was something wrong with his bicycle, too, and every so often the chain came off. So Turing carried a can of oil and a rag to clean up after he replaced it.

Eventually, becoming tired of putting the chain back on, he decided to tackle the problem rationally. He started counting how many times the pedals revolved between one loss of the chain and the next. This number was remarkably constant. Comparing it to the number of links in the bicycle chain and the number of spokes in the back wheel, he deduced that the chain fell off whenever both chain and wheel were in some particular configuration. He then kept a running count to warn him when the chain was about to come off, and carried out a manoeuvre that kept it on. He no longer needed to carry oil and rag. Eventually he discovered that a slightly bent spoke was coming into contact with a damaged link.

It was a triumph of rationality, but anyone else would have taken the bike to a cycle shop, where the fault would quickly have been found. On the other hand, by *not* doing that, Turing saved the cost of a repair – and made sure that no one else could ride his bike. As in so many other things, he had his reasons; they were just different from everyone else's.

Alan Turing's father Julius was a member of the Indian Civil Service. His mother Ethel (*née* Stoney) was the daughter of the chief engineer of the Madras Railways. The couple wanted their children to be brought up in England, so they moved to London. Alan was the second of two sons. When he was six, he went to school in the coastal town of St Leonard's, where the headmistress quickly realised he was unusually bright.

When he was 13, he attended Sherborne school, an independent 'public' school, the quaint English term for a private fee-paying school frequented mainly by children of the rich. Like most such, the school emphasised classics. Turing had bad handwriting, was poor at English, and even in his favourite subject of mathematics he preferred his own answers to those required by the teachers. Either despite or because of this, he won all the mathematics prizes. He also enjoyed chemistry, but again preferred to plough his own furrow. His headmaster wrote: 'If he is to be solely a scientific specialist, he is wasting his time at a public school.'

Too true.

The school was unaware that in his spare time Turing was reading about relativity from Einstein's papers, and quantum theory from Arthur Eddington's *The Nature of the Physical World*. In 1928 he became close friends with Christopher Morcom, a student one year higher, and they shared an interest in science. But within two years, Morcom was dead. Turing was devastated, but he soldiered on doggedly, winning a place to study mathematics at King's College, Cambridge. He continued to read textbooks that were well ahead of, or outside, the undergraduate curriculum. He graduated in 1934.

Turing was incorrigibly scruffy. Even when he wore a suit it was seldom pressed. He is said to have tied up his trousers using a necktie, or sometimes string. His laugh was a loud bray. He had a speech impediment, not so much a stammer as a sudden pause when he would say *ah-ah-ah-ah-ah...* when searching his mind for an appropriate word. He wasn't fastidious about shaving and suffered from 'five o'clock shadow'. He is often portrayed as a nervous, socially inept nerd, but he was actually quite popular, and a good mixer. His apparent eccentricities largely stemmed from the originality not just of what he thought about, but *how* he thought. When working on a problem, Turing found angles that no one else knew existed.

A year later he was taking a postgraduate course on the foundations of mathematics from Max Newman, where he learned about the Hilbert programme and its refutation by Gödel. Turing realised that Gödel's undecidability theorem was really about algorithms. A question is decidable if there exists an algorithm to answer it. You can prove that, for a given problem, by finding one. Undecidability is deeper and more difficult: you must prove that no such algorithm exists. It's hopeless to attempt that unless you have a precise definition of what an algorithm is. Gödel had in effect dealt with this issue by thinking of an algorithm as a proof within an axiomatic system. Turing started thinking about how to formalise algorithms in general.

In 1935 he became a Fellow of King's College, for his independent discovery of the central limit theorem in probability, which provides some rationale for the widespread use of the 'bell curve', or normal distribution, in statistical inference. But in 1936 his thoughts about

Gödel's theorems came to the fore, with the publication of his seminal paper 'On computable numbers, with an application to the *Entscheidungsproblem*' (decision problem). In it, he proved an undecidability theorem for a formal model of computation, now called a Turing machine. He proved that no algorithm can decide in advance whether a computation will stop with an answer. His proof is simpler than Gödel's, although both require preliminary manoeuvres to set up the context.

Although we speak of a Turing *machine*, the name refers to an abstract mathematical model representing an idealised machine. Turing called it an a-machine – 'a' for 'automatic'. It can be thought of as a strip of tape divided into adjacent cells, which are either empty or contain a symbol. The tape is the machine's memory, and its length is unlimited but finite. If you get to the end, add some more cells. A head, positioned over an initial cell, reads the symbol in that cell. It then consults a table of instructions (program supplied by the user), writes a symbol in the cell (overwriting anything already there), and moves the tape one cell sideways. Then, depending on the table and symbol, the machine either stops, or obeys the instructions in the table for the symbol in the cell it has moved to.

There are many variants, but all are equivalent in the sense that they can compute the same things. In fact, this rudimentary machine can in principle compute anything that a digital computer, however fast and advanced, can compute. For example, a Turing machine using the symbols 0–9 and perhaps a few others can be programmed to calculate the digits of π to any specified number of decimal places, writing them in successive cells of the tape, and finally stopping. This level of generality may seem surprising for such a simple device, but the intricacy of the computation is inherent in the table of instructions, which can be very complicated, just as the actions of a computer are inherent in the software it's running. However, the simplicity of a Turing machine also makes it very slow, in the sense that even a simple computation involves a gigantic number of steps. It's not practical, but its simplicity makes it well suited to theoretical questions about limits to computation.

Turing's first important theorem proves the existence of a *universal* Turing machine, which can simulate any specific one. The program of the specific machine is encoded on the universal machine's tape, before the computation starts. The table of instructions tells the universal machine how to decode these symbols into instructions, and carry them out. The universal machine's architecture is an important step closer to a real computer, with a program stored in memory. We don't build a new computer for each problem, with a hard-wired program – except for some very special applications.

His second important theorem does a Gödel, proving that the halting problem for Turing machines is undecidable. This problem asks for an algorithm that can decide, given the program for a Turing machine, whether the machine will (eventually) stop with an answer, or continue for ever. Turing's proof that no such algorithm exists – that the halting problem is undecidable – assumes that it does, and then applies the resulting machine to its own program. However, this is cunningly transformed so that the simulation halts if and only if the original machine doesn't. This leads to a contradiction: if the simulation stops, then it doesn't; if it doesn't, then it does. We saw that Gödel's proof ultimately encodes a statement of the form 'this statement is false'. Turing's, which is simpler, is more like a card bearing on its two sides the messages:

The statement on the other side of this card is true.
The statement on the other side of this card is false.

Each statement, in two steps, implies its own negation.

Turing submitted his paper to the *Proceedings of the London Mathematical Society*, not realising that, a few weeks earlier, the American mathematical logician Alonzo Church had published 'An unsolvable problem in elementary number theory' in the *American Journal of Mathematics*. This provided yet another alternative to Gödel's proof that arithmetic is undecidable. Church's proof was very complicated, but he published first. Newman persuaded the journal to publish Turing's paper anyway, because it was much simpler, both conceptually and structurally. Turing revised it to cite Church's paper, and it appeared in 1937. The tale had a happy ending, though, because Turing then went to Princeton to take a PhD under Church. His thesis was published in 1939 as *Systems of Logic Based on Ordinals*.

✳

Not an auspicious year, 1939 marked the start of World War II. Realising that war was likely, and knowing that modern warfare relied heavily on cryptography – secret codes – the head of the Secret Intelligence Service (SIS or MI6) had bought a property suitable for use as a cipher school. Bletchley Park consisted of a mansion built in a strange mixture of architectural styles, standing in 235 hectares of grounds. The house had been scheduled for demolition, to build a housing estate. It still exists, along with its outbuildings, including some of the wartime huts, and Bletchley Park is now a tourist attraction themed around its wartime codebreakers.

Commander Alastair Dennison, head of operations of the Government Code and Cypher School (GC&CS), moved his top cryptanalysts – codebreakers – to Bletchley Park. They included chess players, crossword solvers, and linguists; one was an expert on Egyptian papyri. Seeking to expand their number, he sought out 'men of the professor type'. The Axis forces were increasingly using machines to encrypt messages, based on complex systems of rotating wheels and daily settings created by plug-in wires. Advanced technical knowledge was therefore required too, and that meant mathematicians. Several joined the team, Newman and Turing among them. They worked in strictest secrecy, supported by clerical staff and administrators. At its peak in early 1945, Bletchley Park had a staff of 10,000.

The main machines used by the Axis forces were the Enigma and Lorenz ciphers. Both cipher systems were thought to be unbreakable, but the mathematical structure of the encryption algorithm had subtle weaknesses. These were exacerbated when users broke the rules and took shortcuts, such as using the same settings on consecutive days, sending the same message twice, or starting messages with standard words and phrases. Turing was a key figure in the team that was trying to break Enigma, working under Dilly Knox of GC&CS. In 1939 the Poles managed to get their hands on a working Enigma machine, and told the British how it worked – how the rotors were wired up. The Polish cryptanalysts also developed methods for breaking the Enigma code, based on the German habit of preceding code messages with a short piece of text allowing the operator to test the machine. For example, a message that continued a previous one would often start with FORT (*Fortsetzung*, 'continuation'), followed by the time the first message was sent, repeated twice and bracketed by the letter Y. The Polish cryptanalysts invented a machine, the *bomba*, to speed things up.

Turing and Knox, realising that the Germans would probably eliminate this flaw, sought more robust methods of decryption, and decided they also needed a machine, which they named the 'bombe'. Turing drew up the specifications for the bombe, which would implement the same general technique of crib-based decryption. This is a method that can be tried when the plaintext version of some part of a message can be guessed – such as the FORT segment. Typical texts of this kind were the German versions of 'nothing to report' and 'weather survey [time]'. Amazingly, Field Marshall Erwin Rommel's quartermaster started every message to him with the identical formal opening phrases.

Turing's design for the bombe was turned into hardware by an engineer named Harold Keen, who worked for the British Tabulating Machine Company (a sort of British IBM). The bombe's task was to use high-speed trial and error to identify some of the basic settings of the Enigma machine, which were (usually) changed every day. It examined each possibility in turn, seeking a contradiction. If it found one, it went on to the next possibility, running through all 17,576 combinations until it hit something plausible. At that point it stopped, and the settings could be read off. Turing improved the process with some statistical analysis. He also tackled the more difficult version of Enigma used by the German Navy. In 1942 he was seconded to the British Joint Staff Mission in Washington DC to advise the Americans on bombes and their uses. His techniques cut the number of machines required from 336 to 96, speeding up their implementation.

The ability to decrypt Axis communications caused a strategic problem: if the enemy realised the Allies could do that, they would tighten up their procedures. So even when the Allies knew the enemy's intentions, any action to defeat them had to be indirect and infrequent. Used with cunning and much deception, the Allies' ability to decrypt enemy code messages helped them win many major engagements, notably the Battle of the Atlantic. The efforts of Turing and his colleagues probably shortened the war by four years.

After the end of the war, it turned out that the German cryptanalysts were aware that the Enigma code could in principle be broken. They just hadn't believed anyone would go to the immense amount of effort required to do it.

The cryptographic work was intense and sustained, but life at Bletchley Park had its lighter moments. Turing relaxed by playing sports and chess, and by socialising with colleagues during the limited time allotted for that purpose. In 1941 he became increasingly friendly with Joan Clarke, a brilliant female mathematician who had abandoned her studies for Part III of the Mathematical Tripos at Cambridge to join the team at Bletchley Park. They went to the cinema together and generally enjoyed each other's company. The relationship grew ever closer, and eventually Turing proposed marriage. Joan immediately accepted.

He had made her aware of his homosexual tendencies, but this failed to discourage her, possibly because they had enough in common – chess, mathematics, cryptography … Few men in those days would have wanted a mathematical prodigy as a wife, but this wasn't an issue for Turing. Neither was his homosexuality, at least, not to begin with. At that time, respectability was more important to many people than sexual orientation, and a wife's main role was seen to be that of housekeeper. Turing did, however, allow Joan to believe that his homosexuality was only a tendency, not actual sexual activity. They met each other's parents, without any problems, and Turing bought her an engagement ring. Joan didn't wear it to work, and among their colleagues only Shaun Wylie officially knew they were engaged, but the others suspected something of the kind.

As the year went by, Turing began to have second thoughts. They spent a week's leave walking and cycling in North Wales, but the holiday ran into problems with a hotel booking, and Turing had forgotten to arrange a temporary ration card to buy food. Shortly after their return, he decided that marriage would not be in the interests of either of them, and the engagement was broken off. He managed to do this without making Joan feel rejected, and they continued to work together, though less frequently than before.

Turing was a high-class athlete, specialising in long-distance running, where his lack of speed was more than compensated for by unusual stamina. As a fellow at King's College he frequently ran the 50-kilometre round trip from Cambridge to Ely and back, and during the war he would run between London and Bletchley Park for meetings. In 1946 the magazine *Athletics* listed him as the winner of Walton Athletic Club's three-mile title, in 15 minutes 37·8 seconds – a respectable time but nothing out of the ordinary. He did cross-country running, and the following year came third in the Kent 20-mile road race in a time of 2 hours, 6 minutes, and 18 seconds – four minutes behind the winner; then fifth in an AAA marathon in a time of 2 hours, 46 minutes, and 3 seconds. The Club's secretary wrote: 'We heard him rather than saw him. He made a terrible grunting noise when he was running, but before we could say anything to him, he was past us like a shot out of a gun.' In 1948, when Britain hosted the Olympic Games, Turing came fifth in the trials for the British marathon team. The gold medallist's time was only 11 minutes less than Turing's personal best.

After the war Turing moved to London, and worked on the design of one of the first computers, ACE (Automatic Computing Engine) at the National Physical Laboratory. Early in 1946 he gave a presentation on the design of a stored-program computer – far more detailed than the American mathematician John von Neumann's slightly earlier design for EDVAC (Electronic Discrete Variable Automatic Computer). The ACE project was slowed down by official secrecy about Bletchley Park, so Turing went back to Cambridge for a year, writing an unpublished article about machine intelligence, his next great theme. In 1948 he became Deputy Director of the Computing Machine Laboratory at the University of Manchester, along with the post of reader (roughly equivalent to associate professor in the USA). In 1950 he wrote 'Computing machinery and intelligence', proposing the now famous Turing test for intelligence in a machine; basically, you can have a long conversation with it on any topic you wish and you won't be able to tell you're not talking to a human (as long as you can't see it). Although controversial, this was the first serious proposal along such lines. He also started work on a chess-playing program for a hypothetical machine. He tried to run it on a Ferranti Mark 1, but the memory was too small, so he simulated the program by hand. The machine lost. But only 46 years later, IBM's Deep Blue beat chess grandmaster Gary Kasparov, and a year later an updated program won a series against him $3\frac{1}{2} - 2\frac{1}{2}$. Turing was just ahead of his time.

From 1952 to 1954 he turned to mathematical biology, especially morphogenesis – the creation of form and patterns in plants and animals. He worked on phyllotaxis, the remarkable tendency of plant structures to involve Fibonacci numbers 2, 3, 5, 8, 13, and so on, each being the sum of the previous two. His biggest contribution was to write down differential equations that model pattern formation. The underlying idea was that chemicals called morphogens lay down a cryptic 'pre-pattern' in the embryo, which acts as a template for the patterns of coloured pigment that appear as the creature grows. The pre-pattern is created by a combination of chemical reactions and diffusion, in which molecules spread from cell to cell. The mathematics of such systems shows that they can form patterns by a mechanism known as symmetry-breaking, which occurs if the uniform state (all chemical concentrations the same everywhere) becomes unstable. Turing explained this effect: 'If a rod is hanging from a point a little above its centre of gravity it will be in stable equilibrium. If, however, a mouse climbs up the rod, the equilibrium eventually becomes unstable and the rod starts to swing.' A swinging rod is in a less symmetric state than one hanging vertically.

However, biologists came to prefer a different approach to the growth and form of the embryo, known as positional information. Here an animal's body is thought of as a kind of map, and its DNA acts as an instruction book. The cells of the developing organism look at the map to find out where they are, and then at the book to find out what they should do at that location. Coordinates on the map are supplied by chemical gradients: for example, a chemical might be highly concentrated near the back of the animal and gradually fade away towards the front. By 'measuring' the concentration, a cell can work out where it is. Evidence supporting the theory of positional information came from transplant experiments, in which tissue in a growing embryo is moved to a new location. For example, a mouse embryo starts to develop a kind of striped pattern that eventually becomes the digits that make up its paws. Transplanting some of the tissue provides insight into the chemical signals it receives from surrounding cells. The experimental results were consistent with the theory of positional information, and were widely interpreted as confirming it.

However, in December 2012 a team of researchers led by Rushikesh Sheth carried out more complex experiments. They showed that a particular set of genes affects the number of digits that the mouse develops. As the effect of these genes decreases, the mouse grows more digits than usual – like a human with six or seven fingers instead of five. Their results are incompatible with the theory of positional information and chemical gradients, but make complete sense in terms of Turing's reaction–diffusion approach. In the same year a group under Jeremy Green showed that ridge patterns inside a mouse's mouth are controlled by a Turing process.[12] The morphogens involved are Fibroblast Growth Factor and Sonic Hedgehog, so called because laboratory fruit flies lacking the fly version have extra bristles on their bodies.

Turing was gay, and in 1952, when he began a relationship with an unemployed 19-year-old named Arnold Murray, active homosexuality was illegal. A burglary at Turing's home, by someone who knew Murray, led to an investigation by the police which uncovered the homosexual relationship. Turing and Murray were charged with gross indecency. On solicitor's advice, Turing pleaded guilty, while Murray received a conditional discharge. Turing was given a choice of imprisonment, or probation accompanied by hormone treatment with a synthetic oestrogen. In *Prof: Alan Turing Decoded* his nephew Dermot Turing, a lawyer, argues that the sentencing was 'procedurally flawed, partly illegal, and ineffective'. In particular, others prosecuted at the same time were treated more leniently, and the person with whom he committed the offence effectively went unpunished. Turing chose probation and hormone treatment, predicting: 'No doubt I shall emerge from it all a different man, but quite who I've not found out.' He was right. He became impotent and developed breasts.

The conviction seems to have been driven by official panic. The recent discovery that Guy Burgess and Donald Maclean were KGB double agents had exacerbated fears about Soviet agents recruiting homosexuals as spies by threatening to expose them. The Government Communications Headquarters (GCHQ), which had developed from GC&CS, promptly removed Turing's security clearance, and the United States refused him entry. So Alan Turing, a man whose mathematical genius had shortened World War II by years (for which he was awarded an OBE – he deserved a knighthood) became *persona non grata* on both sides of the Atlantic.

In June 1954 his housekeeper found his dead body. The post-mortem reported cyanide poisoning as the cause. There was a partially eaten apple beside him, and this was assumed to be the source of the cyanide, although – bizarrely – it wasn't tested for the substance. The Coroner's verdict was suicide. Another possibility seems to have been ignored. Turing might have inhaled cyanide fumes from an electroplating experiment in his spare room. He usually ate an apple before going to bed, and often left it half eaten. He had shown no signs of being depressed by his hormone treatment, and had just made a list of tasks he needed to perform when he got back to the office after a public holiday. So his death could have been accidental.

In 2009, after a campaign on the internet, the Prime Minister Gordon Brown gave a public apology for Turing's 'appalling' treatment. Continuation of the campaign led to a posthumous pardon in 2013 by Queen Elizabeth II. In 2016 the British government announced that all gay and bisexual men convicted of now-abolished sexual offences would be pardoned, in an amendment to the Policing and Crimes Bill informally known as 'Turing Law'. However, some campaigners continue to insist on an apology, not a pardon, on the grounds that a pardon implies commission of an offence.

24

Father of Fractals
Benoit Mandelbrot

Benoit B. Mandelbrot
Born: Warsaw, Poland, 20 November 1924
Died: Cambridge MA, USA, 14 October 2010

DISRUPTION CAUSED BY WORLD WAR II delayed the 1944 entrance examinations for Paris's two great educational institutions, the École Normale Supérieure and the École Polytechnique, by six months. The examinations lasted a month and were extremely difficult, but young Benoit Mandelbrot completed both. One of his teachers discovered that out of all the candidates, just one had answered one particularly difficult mathematics question. He guessed it must have been Mandelbrot, and on asking, discovered he was right. The teacher confided that he'd found the problem impossible himself, because of a 'truly horrible triple integral' which lay at the heart of the calculation.

Mandelbrot laughed. 'It's very simple.' Then he explained that the integral was actually the volume of a sphere, in disguise. If you used the right coordinate system, it was obvious. And everyone knew the formula for the volume of a sphere. That's all there was to it. Once you saw the trick ... Mandelbrot was obviously right. Shocked, the teacher wandered off, muttering 'Of course, of course.' Why hadn't he spotted that himself?

Because he'd been thinking symbolically, not geometrically.

Mandelbrot was a natural geometer, with a strong visual intuition. After a difficult childhood, as a Jew in occupied France in constant danger of being arrested by the Nazis, and most likely ending up in a death camp, he carved out for himself an unorthodox but highly creative mathematical career, the core of which was spent as a Fellow at IBM's Thomas J. Watson Laboratories in Yorktown Heights, New York state. There, he produced a series of articles on topics ranging from the frequencies of words in languages to the flood levels of rivers. Then, in a burst of inspiration, he synthesised the bulk of these diverse and curious researches into a single geometric concept: that of a fractal.

The traditional shapes of mathematics, such as spheres, cones, or cylinders, have a very simple form. The closer you look, the smoother and flatter they appear to be. The overall detail disappears, and what's left looks much like a featureless plane. Fractals are different. A fractal has detailed structure on any scale of magnification. It's infinitely wiggly. 'Clouds are not spheres,' Mandelbrot wrote, 'mountains are not cones, coastlines are not circles and bark is not smooth, nor does lightning travel in a straight line.' Fractals capture aspects of nature that the traditional structures of mathematical physics don't. They've led to fundamental changes in how scientists model the real world, with applications to physics, astronomy, biology, geology, linguistics, global finance, and many other areas. They also have deep pure mathematical features, and strong links to chaotic dynamics.

Fractals are one of several areas of mathematics that, while not entirely new, took off during the second half of the twentieth century, and changed the relationship between mathematics and its applications by providing new methods and viewpoints. The roots of fractal geometry can be traced back to the search for logical rigour in analysis, leading to the invention around 1900 of a variety of 'pathological curves' whose main role was to show that naive intuitive arguments can go wrong. For instance, Hilbert defined a curve that passes through every point inside a square – not just comes close, but hits every point exactly. It's called a space-filling curve, for obvious reasons, and it cautions us to take care when thinking about the concept of dimension. A continuous transformation can *increase* the dimension of a space, here from 1 to 2. Other examples are Helge von Koch's snowflake curve, which has infinite length but encloses finite area, and Wacław Sierpiński's gasket, a curve that crosses itself at every point.

However, those early works had little significance outside specialist areas, and were mainly seen as isolated curiosities. In order for a subject area to 'arrive', someone has to pull the pieces together, understand their underlying unity, formulate the required concepts in sufficient generality, and then go out and sell the ideas to the world. Mandelbrot, though by no means a mathematician in the orthodox sense, had the vision and the tenacity to do just that.

Benoit was born into an academic family of Lithuanian Jews, in Warsaw, between the wars. His mother Bella (*née* Lurie) was a dentist. His father Karl Mandelbrojt, who had had no formal education, made and sold clothes, but his side of the family largely consisted of scholars, going back for generations, so Benoit was raised in an academic tradition. Karl had a younger brother Szolem, who later became a distinguished mathematician. Because his mother had already lost one child through an epidemic, she kept Benoit out of school for several years, to avoid the possibility of infection. Another uncle, Loterman, taught him at home, but he wasn't a very effective teacher. Benoit learned to play chess, and listened to classical myths and stories, but did little else. He didn't even learn the alphabet or his multiplication tables. He did, however, develop an aptitude for visual thinking. His chess moves were dictated more by the shape of the game – the pattern of pieces on the board. He loved maps, a predilection that he probably got from his father, who was an avid map collector. They were hung on all the walls. He also read anything he could lay his hands on.

The family left Poland in 1936 as economic and political refugees. His mother had been unable to continue as a medic and his father's business had collapsed. They moved to Paris, where his father had a sister. Later Mandelbrot credited her with saving their lives and helping them ward off depression.

Szolem Mandebrojt was moving up in the mathematical world, and when Benoit was five his uncle became a professor at the University of Clermont-Ferrand. Eight years later he advanced to the position of professor of mathematics at the Collège de France, Paris. Mandelbrot, impressed, began thinking about a career in mathematics himself, although his father disapproved of such an impractical occupation.

When Mandelbrot was in his teens, uncle Szolem took charge of his education. He went to the Lycée Rolin in Paris. But occupied France was a bad place and time to be a Jew, and his childhood was marked by poverty and the constant threat of violence or death. In 1940, the family fled again, this time to the tiny town of Tulle in southern France where his uncle had a country house. Then the Nazis occupied southern France as well, and Mandelbrot spent the next eighteen months evading capture. He described this period of his life in bleak terms:[13]

> For some months I was in Périgueux as apprentice toolmaker on the railroads. For later use in peacetime, the experience was better than another wartime stint as horse groom, but I did not look or talk like an apprentice or groom and, at one point, narrowly escaped execution or deportation. Some good friends eventually arranged for admission to the Lycée du Parc, in Lyons. While much of the world was in turmoil, it was almost business as usual in a class preparing for the feared examinations of the French elite universities called 'Grandes Écoles'. The few months that followed in Lyons were

among the most important of my life. Stark poverty and deep fear of the German boss of the city (we later discovered his name to be Klaus Barbie) tied me to my desk for most of the time.

Barbie was a *Hauptsturmführer* in the dreaded *Schutzstaffel* (SS, literally 'protection squad') and a member of the Gestapo (secret police). He became known as the Butcher of Lyon for torturing French prisoners in person. After the war he fled to Bolivia, but was extradited to France in 1983 and imprisoned for crimes against humanity.

At Lyon in 1944, studying mathematics, Mandelbrot discovered that he had a high degree of visual intuition. When his teacher posed some difficult problem in symbolic form, such as an equation, he instantly transformed it into a geometric equivalent, which was usually much easier to solve. He was admitted to the École Normale Supérieure in Paris, to study mathematics. However, the mathematical style that was practised there was very much that of the Bourbaki school – abstract, general, focused on pure mathematics. His uncle had a similar mathematical philosophy, and had been an early member of Bourbaki before the group began its systematic revision of mathematics on rigorous abstract lines. This formal style of mathematical thinking, without pictures or concrete applications, did not appeal to Mandelbrot. After a few days at the École Normale, he decided he was in the wrong place, and resigned. Instead, he took up a place in the more practically oriented École Polytechnique (he'd already passed the entrance examination for that, along with the exam for the École Normale). Here he had much more freedom to study different disciplines.

His uncle continued to push him towards more abstract mathematics, and suggested that Mandelbrot should choose a PhD topic related to work of Gaston Julia on complex functions, which had been published in 1917. This suggestion did not appeal. When accepting the Wolf Prize, Mandelbrot later wrote:[14]

> My uncle's beloved Taylor and Fourier series had started centuries ago in the context of physics, but in the 20th century developed into a field self-described as 'fine' or 'hard' mathematical analysis. In my uncle's theorems, the assumptions could take pages. The distinctions he enjoyed were so elusive that no condition was both necessary and

sufficient. The long pedigree of the issues, for him a source of pride, was for the younger me a source of aversion.

One day, still seeking a topic, Mandelbrot asked Szolem for something to read on the metro. His uncle remembered having thrown an article into the wastebasket, and fished it out, saying that it was 'crazy, but you like crazy things'. It was a review of a book by the linguist George Zipf, about a statistical property that was common to all languages. No one seemed to understand what it was about, but Mandelbrot decided on the spot that he would explain this property, now called Zipf's law. He made some progress, as we'll soon see.

From 1945 to 1947 Mandelbrot studied under Paul Lévy and Gaston Julia at the École Polytechnique, and then went to the California Institute of Technology, obtaining a master's degree in aeronautics. He then went back to France, getting a PhD degree in 1952. He was also employed at the Centre National de la Recherche Scientifique. He spent a year at the Institute for Advanced Study in Princeton, New Jersey, under the sponsorship of John von Neumann. In 1955 he married Aliette Kagan and moved to Geneva. After several visits to the USA the Mandelbrots moved there permanently in 1958, and Benoit worked as an IBM researcher in Yorktown Heights. He remained at IBM for 35 years, becoming an IBM Fellow and then a Fellow Emeritus. He received numerous awards, including the Légion d'Honneur (1989), the Wolf Prize (1993), and the Japan Prize (2003). His books include *Fractals: Form, Chance, and Dimension* (1977) and *The Fractal Geometry of Nature* (1982). He died of cancer in 2010.

The work on Zipf's law set the pattern of Mandelbrot's future career, which for a long time seemed to be a series of apparently unrelated investigations of strange statistical patterns, hopping like a butterfly from one weird flower to another. Only when he was at IBM did it all start to come together.

Zipf's law introduced him to a simple but useful (and underestimated) idea in statistics, that of a power-law relationship.

In one standard compilation of American English, the three most common words are:

the, occurring 7 per cent of the time,
of, occurring 3.5 per cent of the time,
and, occurring 2.8 per cent of the time.

Zipf's law states that the nth word (ranked by the frequency with which it occurs) is the frequency of the first word, divided by n. Here $7/2 = 3{\cdot}5$ and $7/3 = 2{\cdot}3$. The latter figure is lower than observed, but the law isn't perfect, it just quantifies a general tendency. Here the frequency of the nth word in the ranking is proportional to $1/n$, which we can write as n^{-1}. Other examples show similar patterns, but with a power that differs from -1. For example, in 1913 Felix Auerbach noticed that the size distribution of cities follows a similar law but with the power $n^{-1.07}$. In general, if the rank-n item has frequency proportional to n^c, for some constant c, we speak of a cth power law.

Classical statistics pays little attention to power-law distributions, focusing instead on the normal distribution (or bell curve), for a variety of reasons, some good. But nature often seems to use power-law distributions instead. Laws like Zipf's apply to the populations of cities, the numbers of people watching a selection of TV shows, and how much money people earn. The reasons for this behaviour are still not fully understood, but Mandelbrot made a start in his thesis,[15] and Wentian Li has offered a statistical explanation: in a language where each letter in the alphabet (plus a space character to separate words) appears with the same frequency, the distribution of words by rank obeys an approximation to Zipf's law. Vitold Belevitch proved that the same goes for a variety of statistical distributions. Zipf's own explanation was that languages evolve over time to provide optimal understanding for the least effort (in speaking or listening), and the power -1 emerges from this principle.

Subsequently, Mandelbrot published papers on the distribution of wealth, the stock market, thermodynamics, psycholinguistics, the lengths of coastlines, fluid turbulence, population demographics, the structure of the universe, the areas of islands, the statistics of river networks, percolation, polymers, Brownian motion, geophysics, random noise, and other disparate topics. It all seemed a bit disjointed. But in 1975 everything came together in a flash of insight: there was a common underlying theme to almost all of his work. And it was geometric.

The geometry of natural processes seldom follows the standard mathematical models of spheres, cones, cylinders, and other smooth surfaces. Mountains are jagged and irregular. Clouds are fluffy with bulges and wisps. Trees branch repeatedly, from trunk to bough to twig. Ferns have fronds that look like a lot of smaller fronds strung together in opposite pairs. Under a microscope, soot is a lot of tiny particles clumped together, with gaps and voids. They're all a long way from the smooth rotundity of a sphere. Nature abhors a straight line, and she's not too keen on much else from Euclid and the calculus texts. Mandelbrot coined a name for this type of structure: *fractal*. And he energetically and enthusiastically promoted the use of fractals in science, to model many of nature's irregular structures.

'Model' is a key word here. The Earth may appear roughly spherical – ellipsoidal if you want more accuracy – and those shapes have helped physicists and astronomers understand such things as tides and the tilt of the planet's axis, but the mathematical objects are models, not reality itself. They capture some features of the natural world in an idealised form, simple enough for human brains to analyse. But the surface of the Earth is rough and irregular: the map is not the territory. Nor should it be. A map of Australia can be folded and put in your pocket, ready for use when needed, but you can't do that with Australia itself. A map should be simpler than the territory, but provide useful information about it. A mathematical sphere is perfectly smooth no matter how much you magnify it, but reality turns into quantum particles at the atomic level. This is irrelevant to a planet's gravitational field, however, so it can and should be ignored in that context. Water can profitably be modelled as an infinitely divisible continuum, even though real water becomes discrete when you get to the level of molecules.

It's the same with fractals. A mathematical fractal isn't just a random shape. It has detailed structure on all scales of magnification. Often, it has virtually *the same* structure on all scales. Such a shape is said to be self-similar. In a fractal model of a fern, each frond is made of smaller fronds, which in turn are made of even smaller ones, and this process *never stops*. In a real fern, it stops after four or five stages, at most. Nevertheless, the fractal is a better model than, say, a triangle. Just as an ellipsoid can be a better model of the Earth than a sphere.

Mandelbrot was very aware of the prominent role of Polish mathematicians in the prehistory of fractals, a highly abstract approach to analysis, geometry, and topology developed by a small coterie of mathematicians, many of whom met regularly in the Scottish Café in Lvov (Lwów, now Lviv). They included Stefan Banach, who founded functional analysis, and Stanisław Ulam, who was heavily involved in the Manhattan Project to build an atomic bomb, and came up with the main idea for the hydrogen bomb. Wacław Sierpiński, at Warsaw University, was of a like mind, and he invented a shape that was 'simultaneously Cantorian and Jordanian, of which every point is a point of ramification'. That is, a continuous curve that crosses itself at every point.

The first few stages in constructing a Sierpiński gasket

Later Mandelbrot jokingly called this shape the Sierpiński gasket because of its resemblance to the many-holed seal that joins the cylinder head of a car to the engine. Recall that the Sierpiński gasket is one of a small zoo of examples that came into being in the early twentieth century, collectively known as pathological curves – although they're not pathological to nature, or even to mathematics: they just seemed that way to the mathematicians of that period. Patterns like the gasket appear on seashells. Anyway, the gasket can be constructed by an iterative procedure, applied to an equilateral triangle. Divide it into four congruent equilateral triangles, each half the size. Delete the central triangle, which is upside down. Apply the same process to the three remaining triangles, and repeat indefinitely. The gasket is what remains when all the upside-down triangles, but not their edges, are deleted.

Mandelbrot took inspiration from such curves, now seen as early fractals. Later he found this amusing:[16]

> My uncle left for France aged about twenty, a refugee driven by an ideology that was not political or economic but purely intellectual. He was repelled by the 'Polish mathematics', then being built up as a militantly abstract field by Wacław Sierpiński (1882–1969). By profound irony, whose work was to become a fertile hunting ground when, much later, I looked for tools to build fractal geometry? Sierpiński! Fleeing [Sierpiński's] ideology, my uncle joined the heirs of Poincaré who ruled Paris in the 1920s. My parents were not ideological but economic and political refugees; their joining my uncle in France later saved our lives. I never met Sierpiński but his (unwitting) influence on my family had no equal.

A few pure mathematicians, following up such notions, discovered that the degree of roughness of a fractal can be characterised by a number, which they called its 'dimension' because it agrees with the usual dimension for standard shapes like a line, a square with its interior, or a solid cube, which respectively have dimensions 1, 2, and 3. However, the dimension of a fractal need not be a whole number, so the interpretation 'how many independent directions' no longer applies. Instead, what matters is how the shape behaves under magnification.

If you make a line twice the size, its length multiplies by 2. Doubling a square multiplies its area by 4, and doubling a cube multiplies its volume by 8. These numbers are 2^1, 2^2, and 2^3: that is, 2 raised to the power of the dimension. If a gasket is doubled in size, it can be split into three copies of the original. So 2 raised to the power of the dimension ought to equal 3. The dimension is therefore log 3 /log 2, which is about 1·585. A more general definition, not confined to self-similar fractals, is called the Hausdorff–Besicovich dimension, and a more practical version is called the box dimension. The dimension is useful in applications, and is one way to test fractal models experimentally. In this manner, for example, it has been shown that clouds are well modelled by fractals, with a dimension of a photographic image (projection into the plane, easier to work with and measure) being roughly 1·35.

A final irony illustrates the danger of passing snap value judgements in mathematics. In 1980, seeking new applications of fractal geometry, Mandelbrot took another look at Julia's 1917 paper, the one his uncle had recommended and he had rejected as too abstract. Julia, and another mathematician Pierre Fatou, had analysed the strange behaviour of complex functions under iteration. That is, start with some number, apply the function to that to get a second number, then apply the function to that to get a third number, and so on, indefinitely. They focused on the simplest nontrivial case: quadratic functions of the form $f(z) = z^2+c$ for a complex constant c. The behaviour of this map depends on c in a complicated manner.[17] Julia and Fatou had proved several deep and difficult theorems about this particular iteration process, but it was all symbolic. Mandelbrot wondered what the picture looked like.

Left: The Mandelbrot set. *Right*: Magnification of part of it.

The calculations were far too lengthy to carry out by hand, which is probably why Julia and Fatou hadn't investigated the geometry of the process. But now computers were starting to gain real power, and Mandelbrot was working at IBM. So he programmed a computer to do the sums and draw the picture. It was messy (the printer was running out of ink) and crude, but it revealed a surprise. The complicated dynamics of Julia and Fatou is organised by a single geometric object – and it, or more precisely its boundary, is a fractal. The dimension of the boundary is 2, so it's 'almost space-filling'. We now call this fractal the Mandelbrot set, a name coined by Adrien Douady. As always, there were prediscoveries and closely related work; in particular, Robert Brooks and Peter Matelski drew the same set in 1978. The Mandelbrot set is the source of complex and beautiful computer graphic images. It's also the subject of intense mathematical study, leading to at least two Fields medals.

So the abstract pure-mathematical paper that Mandelbrot had initially rejected turned out to contain an idea that became central to the theory of fractals, which he had developed precisely because of its lack of abstraction and its links with nature. Mathematics is a highly connected whole, in which the abstract and the concrete are linked by subtle chains of logic. Neither philosophy is superior. The big breakthroughs often come by using both.

25

Outside In
William Thurston

William Paul Thurston
Born: Washington DC, 30 October 1946
Died: Rochester NY, 21 August 2012

MATHEMATICIANS LIKE NOTHING BETTER than to talk to other mathematicians – about their work, in the hope of learning of a new idea to help with their latest problem; about the new Thai restaurant that's just opened up on the edge of the campus; about family and mutual friends. They generally do this sitting around in small groups drinking coffee. As Alfréd Rényi once said, 'A mathematician is a machine for turning coffee into theorems.' It's a pun in German, where the word *Satz* means both 'theorem' and '(coffee) grounds'.

These informal discussions often occur in a more formal context – a seminar (a technical lecture for specialists), a colloquium (a less technical lecture ostensibly for professionals or graduate students who may be working in a different area, although sometimes it's hard to tell the difference), a workshop (smallish specialised conference), a sandpit (smaller and less formal), or a conference (bigger and possibly broader). In December 1971 the University of California at Berkeley hosted a seminar on dynamical systems. This had recently become a hot topic because Stephen Smale, Vladimir Arnold, and their colleagues and students in Berkeley and Moscow were continuing where Poincaré left off when he discovered chaos, devising new topological methods to tackle apparently intractable old problems. A dynamical system is anything that develops over time according to specific non-random rules. The rules for a continuous dynamical system are differential equations, which determine the state of the system a tiny instant into the future in terms of its current state. There's an analogous concept of a discrete dynamical system, in which time ticks in isolated instants, 1, 2, 3, ... The seminar speaker presented a breakthrough solution of a problem that boiled down to looking at a finite number of points in the plane. The speaker explained a key trick: how to move any given number of points to new positions, not too far away, so that they didn't stray too far at any stage of the motion. (A few other conditions had to be obeyed too.) This theorem was easy to prove for spaces of three or more dimensions, but now a long-sought proof for two dimensions had, it was claimed, been found. Lots of interesting results in dynamics followed.

At the back of the room sat a shy young graduate student, looking like a hippie with a thick beard and long hair. He stood up, and rather diffidently said he didn't think the proof was correct. Advancing to the blackboard, he drew two pictures, each showing seven points in the plane, and began to use the methods explained in the talk to move the points of the first configuration to the positions in the second one. He drew the paths along which the points were supposed to move, and these started to get in the way of each other, requiring the next path to make a longer excursion to avoid the obstacle, which in turn created an even longer obstacle. As curves sprouted like the heads of the mythical hydra, it became clear that the student was right. Dennis Sullivan, who was present, wrote: 'I had never seen such comprehension and such creative construction of a counterexample done so quickly. This combined with my awe at the sheer complexity of the geometry that emerged.'

The student was William Thurston – 'Bill' to friends and colleagues. There are dozens of similar stories about him. He had a natural intuition for geometry, especially when it got really complicated, and the newly developing geometry of many dimensions – four, five, six, you name it – provided ample scope for the exercise of his astonishing ability to convert technical problems into visual form, and then solve them. He had a knack for seeing through the complexity and uncovering simple underlying principles. He became one of the leading topologists of his age, solved many important problems, and introduced a few key conjectures of his own that defeated even his prodigious talents. Bill Thurston, a truly significant figure of modern pure mathematics, is a worthy representative of this esoteric species.

Ironically, Thurston had poor eyesight. He suffered from congenital strabismus – he was 'cross-eyed', and could not focus both eyes on the same nearby object. This affected his depth perception, so that he found it difficult to work out the shape of a three-dimensional object from a two-dimensional image. His mother Margaret (*neé* Martt) was a proficient seamstress, able to sew patterns so complicated that neither Thurston nor his father Paul could understand them. Paul was an engineer-physicist at Bell Labs with a penchant for hands-on construction of gadgets. And, on one occasion, hands in: he showed young Bill how to boil water with your bare hands. (Use a vacuum pump to lower the boiling point to just above room temperature; then stick your hands in to warm it up.) To help combat Bill's strabismus, Margaret spent hours with him when he was two years old, looking at books full of coloured patterns. His later love of patterns, and his facility as a handyman, probably stem from these early activities.

His early schooling was unusual. New College Florida took a small number of students, selected for outstanding ability, and imposed very few limits on what they studied, or even where they lived. On occasion, Thurston lived in a tent in the woods; at other times he slept in the school buildings, dodging the janitor. The school nearly collapsed after eighteen months when half of its teachers resigned. His university days at Berkeley were more organised, but that too was a turbulent time, with much student opposition to the Vietnam war. Thurston joined a committee that was trying to persuade mathematicians not to accept military funding. He had by then married Rachel Findley, and their first child was born. The birth, Rachel said, was in part intended to make sure Thurston wasn't drafted into the army. Labour coincided with his PhD qualifying examination, and Thurston's performance was erratic, but, as always, original. His PhD thesis was on some special problems in the then-hot topic of foliations, in which a multidimensional space (or manifold) is decomposed into closely fitting 'leaves', like a book decomposes into pages but with less regularity in their arrangement. This topic is related to the topological approach to dynamical systems. The thesis contains several important results, but has never been published. Foliations were Thurston's first main area of research, and he continued working on them at the Institute for Advanced Study in Princeton in 1972–73, and at MIT in 1973–74. Indeed, he solved so many of the basic problems of the area that as far as other mathematicians were concerned, in the end he pretty much killed it off completely.

In 1974 Thurston became a professor at Princeton University (not to be confused with the Institute for Advanced Study, which has no students). A few years later his research moved into one of the most difficult areas of topology, three-dimensional manifolds. These spaces are analogous to surfaces, but have one extra dimension. Their study goes back more than a century to Poincaré (Chapter 18), but until Thurston got involved they had always seemed rather baffling. The topology of higher-dimensional manifolds is curious. The easiest dimensions are one (trivial) and two (surfaces, solved classically). The next easiest turned out to be dimension five or higher, mainly because high-dimensional spaces have a lot of room for performing complicated manoeuvres. Even then, the problems are hard. Harder still are four-dimensional manifolds, and the hardest of all are the three-dimensional ones – enough room for huge complexity; not enough room to simplify it in any straightforward way.

A standard way to construct n-dimensional manifolds is to take little patches of n-dimensional space and prescribe rules for gluing them together. Conceptually, not in reality. In Chapter 18 we saw how this approach works for surfaces and 3-manifolds (page 193). We also encountered a fundamental question in 3-manifold topology, the Poincaré conjecture. This characterises the three-dimensional sphere in terms of a simple topological property: all loops shrink to a point. A standard way to sneak up on such questions is to generalise them to higher-dimensional analogues. Sometimes a more general question is easier; then you read off the special case you started from. Initial progress was encouraging. In 1961 Stephen Smale proved the Poincaré conjecture in all dimensions greater than or equal to 7. Then John Stallings polished off dimension 6 and Christopher Zeeman did dimension 5. Their methods fail for dimensions 3 and 4, and topologists began to wonder whether those dimensions behave differently. Then, in 1982, Michael Freedman found an extremely complicated proof of the four-dimensional Poincaré conjecture using radically different techniques. At this stage topologists had proved the Poincaré conjecture in every dimension except the one Poincaré had originally asked about. Their methods shed no light whatsoever on that final, resistant case.

Enter Thurston, turning the whole subject on its head.

Topology is rubber-sheet geometry, and Poincaré's question was topological. Naturally, everyone had been attacking it using topological methods. Thurston threw away the rubber sheet and wondered whether the problem was really geometric. He didn't solve it, but his ideas inspired a young Russian, Grigori Perelman, to do just that, some years later.

Recall (Chapter 11) that there are three kinds of geometry:

Euclidean, elliptic, and hyperbolic. These are respectively the natural geometry of spaces with zero curvature, constant positive curvature, and constant negative curvature. Thurston began with a curious fact, which looks almost accidental. He revisited the classification of surfaces – sphere, torus, 2-torus, 3-torus, and so on, as in Chapter 18 – asking what kinds of geometry show up. The sphere has constant positive curvature, so its natural geometry is elliptic. One realisation of the torus is the flat torus, a square with opposite edges identified. A square is a flat object in the plane, so its natural geometry is Euclidean, and the gluing rules give the flat torus the same type of geometry as the square. Finally, although this is less obvious, every torus with two or more holes has a natural hyperbolic geometry. Somehow, the flexible topology of surfaces reduces to rigid geometry – and all three possibilities occur.

Of course, surfaces are very special, but Thurston wondered whether something similar works for 3-manifolds. With his amazing intuition for geometry he quickly realised that it couldn't be as simple. Some three-dimensional manifolds, such as the flat torus, are Euclidean. Some, such as the 3-sphere, are elliptic. Some are hyperbolic. But most are none of these things. Undaunted, he investigated why, and found two reasons. First, there are *eight* sensible geometries for 3-manifolds. One, for instance, is analogous to a cylinder: flat in some directions, positively curved in the others. The second obstacle is more serious: many 3-manifolds are still not accounted for. However, what seemed to work was a kind of jigsaw effect. Every 3-manifold seemed to be built from pieces, such that each piece had a natural geometry selected from those eight possibilities. Moreover, the pieces weren't any old piece: they could be chosen to fit together in a fairly stringent way. These ideas led Thurston to state, in 1982, his Geometrisation Conjecture: *every* three-dimensional space can be cut up, in an essentially unique manner, into pieces, each having a natural geometric structure given by one of his eight geometries. The Poincaré Conjecture for 3-manifolds is a simple consequence. But there the matter stuck. The Clay Mathematics Institute made the Poincaré conjecture one of its Millennium Prize problems, with one million dollars as a reward for a proof.

In 2002, Perelman put a preprint on a website called the arXiv ('archive'), about something called the Ricci flow. This idea is related to General Relativity, in which gravity is an effect of the curvature of space-time. Earlier Richard Hamilton had wondered whether the Ricci flow might give a simple proof of the Poincaré conjecture. The idea was to start with a hypothetical 3-manifold such that every closed curve shrinks to a point. Think of it as a curved three-dimensional space in Einstein's sense – an idea that originated in Riemann's Habilitation thesis (Chapter 15).

Now comes the clever bit: try to redistribute the curvature to make it more even.

Imagine trying to iron a shirt. If you're not careful putting it on the ironing-board, it has lots of bumps and ridges. These are regions of high curvature. Elsewhere the shirt is flat – zero curvature. You can try to flatten out the bumps by ironing them, but cloth doesn't stretch or compress terribly well, so either the bumps move somewhere else, or you create wrinkles. A simpler and more effective method, which stops the bumps moving around or reappearing, is to grab the edges of the shirt and stretch it out. Then the natural dynamics of the cloth flattens out the bumps. The Ricci flow does the same sort of thing for a 3-manifold. It redistributes the curvature from regions of high curvature into regions of lower curvature, as if the space is trying to even its curvature out. If everything works nicely, the curvature keeps flowing until it's the same everywhere. Maybe the result is flat, maybe not, but either way its curvature ought to be the same at every point.

Hamilton showed that this idea works in two dimensions: a bumpy surface on which every closed curve shrinks to a point can be ironed out by its Ricci flow until it ends up with constant positive curvature, which implies that it's a sphere. But in three dimensions there are obstacles, and the flow can get stuck where bits of the manifold bunch together and create wrinkles. Perelman found a way to get round this – basically, by cutting off that bit of shirt, ironing it separately, and sewing it back in. His preprint and a follow-up claimed that this method proves both the Poincaré Conjecture and Thurston's Geometrisation Conjecture.

Usually, claims to have solved some huge conjecture initially meet with scepticism. Most mathematicians have found promising proofs of their own, for some difficult problem that interests them, only to discover a subtle mistake. But even early on, there was a general feeling that Perelman might well have cracked it. His method for proving the Poincaré Conjecture looked plausible; the Geometrisation Conjecture was perhaps more troublesome. However, a consensual feeling isn't enough: the proof must be checked. And the ArXiv version – all that existed – left a lot of gaps for readers to fill in, on the grounds that these steps were obvious. Actually, filling in gaps and checking the logic took several years.

Perelman was extraordinarily talented, and what seemed obvious to him wasn't anything like as obvious to those mathematicians who tried to check his proof. To be fair, they hadn't been thinking about the problem the way he had, or for anything like as long as he had, which put them at a disadvantage. He was also a bit reclusive, and as time went by and no one had yet pronounced definitively on what eventually turned out to be epic, groundbreaking work, he became annoyed and disillusioned. By the time his proof was accepted, he'd abandoned mathematics altogether. He refused the million-dollar prize, which was offered to him even though technically he hadn't complied with the conditions because his proof hadn't been published in a recognised journal. He rejected the award of a Fields medal, generally considered the mathematical equivalent of a Nobel Prize, though carrying a much smaller monetary reward. Eventually the Clay Institute used the prize money to set up a short-term position for outstanding young mathematicians at the Institut Henri Poincaré in Paris.

Today many mathematicians use computers, not just for email and the web, not even just for big numerical calculations, but as a tool to help them explore problems in an almost experimental fashion. Indeed, computer-assisted proofs turn up from time to time, often in connection with important problems that have resisted the more traditional methods of pen, paper, and human brainpower. This relaxed attitude to computers is relatively recent; not because mathematicians are stick-in-the-muds who resist new technology, but because computers were previously too limited, both in speed and memory. A serious mathematical problem can stretch even the fastest supercomputer; one recent investigation would have had an output the size of Manhattan if it had been printed out.

Thurston's revival of three-dimensional hyperbolic geometry led him to pioneer the use of computers at the frontiers of geometry. In the late 1980s the National Science Foundation funded a new Geometry Center at the University of Minnesota, which hosted research meetings and public outreach activities. It also advanced the use of computer graphics, and two of its videos achieved considerable fame. They're still available on the web, even though the Center itself is now defunct. The first, *Not Knot*, flies the viewer through various three-dimensional hyperbolic manifolds that Thurston discovered. Its complex and intriguing graphics are so psychedelic that extracts have been used in concerts by the Grateful Dead. *Outside In* is an animation of a remarkable theorem that Smale discovered as a graduate student in 1957. Namely, you can turn a sphere inside out.[18]

Imagine a sphere whose outside is painted gold and inside purple. Of course you can turn it inside out by making a hole and pushing it through the hole, but that's not a topological transformation. The trick is clearly impossible with a real sphere, such as a balloon (though it's not *totally* obvious how to prove this), but mathematically we can allow transformations where the sphere passes through itself, which you can't do with a balloon. Now you could try pushing from opposite sides so that two purple bulges poke through the golden surface, but that leaves an ever-tightening tubular ring of gold. When the ring contracts down to a circle, the surface ceases to be smooth. Smale's theorem says that this can be avoided: there's a transformation such that at all stages the sphere is smoothly embedded in space, though perhaps cutting through itself. For a long time this was a pure existence proof: no one actually knew how to do it. Then various topologists worked out different methods; one, Bernard Morin, had been blind from the age of six. The most elegant and symmetrical method is Thurston's, and that's the star of *Inside Out*.

Thurston had an impact on the public appreciation of mathematics in other ways, too. He wrote about what it's really like to be a mathematician, and how he thought about research problems, giving outsiders an inside view. When Dai Fujiwara, a fashion designer, heard of the eight geometries, he got in touch with Thurston, and their interaction led to an extensive array of women's fashions.

Thurston's contribution to many areas of geometry, ranging from topology to dynamics, is extensive. His work is characterised by a remarkable ability to visualise complex mathematical concepts. When asked for a proof he would usually draw a picture. Thurston's pictures often revealed hidden connections that no one else had noticed. Another characteristic was his attitude to proofs: he often left out details because to him they seemed obvious. When anyone asked him to explain a proof that they didn't understand, he would often invent a different one on the spot and say: 'Perhaps you'll prefer this one.' To Thurston, all mathematics was a single connected whole, and he knew his way round it like other people know their own back garden.

Thurston died in 2012, after surgery for a melanoma in which he lost his right eye. While undergoing treatment, he continued to do research, proving fundamental new results in the discrete dynamics of rational maps of the complex plane. He went to mathematical conferences, and worked to inspire young people about his beloved subject. Whatever the obstacles, he never gave up.

Mathematical People

WHAT, THEN, HAVE WE LEARNED from our significant figures, whose pioneering discoveries opened up new mathematical vistas?

The most obvious message is diversity. Mathematical trailblazers come from all periods of history, all cultures, and all ranks of life. The stories I've selected here span a period of 2500 years. Their protagonists lived in Greece, Egypt, China, Persia, India, Italy, France, Switzerland, Germany, Russia, England, Ireland, and America. Some were born into wealthy families – Fermat, King, Kovalevskaia. Many were middle class. Some were born poor – Gauss, Ramanujan. Some came from academic families – Cardano, Mandelbrot. Some didn't – Gauss and Ramanujan again, Newton, Boole. Some lived in troubled times – Euler, Fourier, Galois, Kovalevskaia, Gödel, Turing. Some were fortunate to have lived in a more stable society, or at least a more stable part of it – Madhava, Fermat, Newton, Thurston. Some were politically active – Fourier, Galois, Kovalevskaia. The first two were imprisoned as a result. Others kept their politics to themselves – Euler, Gauss.

There are some partial patterns. Many grew up in intellectual families. Some were musical. Some were good with their hands, others couldn't fix a bicycle. Many were precocious, showing unusual talent at an early age. Minor coincidences, such as the choice of bedroom wallpaper, a chance conversation, a borrowed book, triggered a life-changing interest in mathematics. Many started out trying to pursue a different career, especially in law or the clergy. Some were encouraged by their proud parents, some were forbidden to study mathematics, some were grudgingly permitted to follow their calling.

Some were eccentric. One was a rogue. A few were crazy. Most were normal, inasmuch as any of us counts as normal. Most married and had families, but some – Newton, Noether – didn't.

Most were male – a cultural bias. Until recently, women were often considered unsuited, by biology and temperament, to mathematics, indeed to any science. Their education, it was said, should be in domestic skills: crochet, not calculus. Their society reinforced this view, and women were often as vocal as men about the unsuitability of mathematics as a womanly pursuit. Even if women wanted to study the subject, they were forbidden to attend lectures, to take examinations, to graduate, and to join the ranks of academe. Our female trailblazers blazed two trails: one through the jungle of mathematics, the other through the jungle of male-dominated society. The second trail made the first even harder. Mathematics is difficult enough when you can get training, books, and time to think. It's almost impossible when you have to battle to obtain any of those things. Despite these obstacles, a few great women mathematicians broke down the barriers, blazing a trail for others to follow. Even today, women are generally under-represented in mathematics and science, but it's no longer socially acceptable to attribute this to differences in ability or mentality, as several prominent men have discovered to their dismay. Nor is there a shred of evidence to support those views.

It's tempting to look for neurological explanations of unusual mathematical talent. In the early days of phrenology, Franz Gall proposed that important abilities are associated with specific regions of the brain, and can be assessed by measuring the shape of the skull. If you're good at mathematics, your head will have a mathematics bump. Phrenology is now seen as pseudoscience, although specific regions of the brain do sometimes play specific roles. Today's obsession with genetics and DNA makes it natural to ask whether there's a 'mathematics gene'. It's hard to see how, because mathematics goes back only a few thousand years, so evolution hasn't had time to select for mathematical ability, any more than it can have selected for the ability to pilot a fighter jet. Mathematical talent presumably exploits other attributes, more conducive to survival – keen vision, a retentive memory, skill at swinging through trees. Sometimes mathematics seems to run in families – the Bernoullis – but mostly it doesn't. Even when it does, the influences are often nurture, not nature: a mathematical uncle, calculus on the wallpaper. Even geneticists are slowly coming to realise that DNA isn't everything.

The pioneering mathematicians do have some generalities in common. They're original, imaginative, and unorthodox. They seek patterns and relish solving hard problems. They pay close attention to logical fine points, but they also indulge in creative leaps of logic, becoming convinced that some line of attack is worth pursuing even when there's little to justify that view. They have strong powers of concentration, yet, as Poincaré urged, they shouldn't be so obsessive that they keep banging their heads against brick walls. They need to give their subconscious minds time to mull everything over. They often have excellent memories, but some – Hilbert, for instance – don't.

They can be lightning calculators, like Gauss. Euler once settled an argument between two other mathematicians, about the fiftieth decimal place in the sum of a complicated series, by doing the sums *in his head*. On the other hand, they can be terrible at arithmetic without any obvious disadvantage. (Most lighting calculators are hopeless at anything more advanced than arithmetic; Gauss, as ever, was an exception.) They have the ability to absorb huge quantities of previous research, distilling its essence and making it their own, but they can also ignore conventional paths altogether. Christopher Zeeman used to say that it was a mistake to read the research literature before starting on a problem, because doing so would slot your mind into the same grooves everyone else was trapped in. Early in his career, the topologist Stephen Smale solved what everyone thought was a truly horrible problem because no one had told him it was hard.

Nearly all mathematicians have a strong intuition, either formal or visual. Here I'm referring to the visual areas of the brain, not to eyesight: Euler's productivity *increased* when he went blind. In *The Psychology of Invention in the Mathematical Field*, Jacques Hadamard asked a number of leading mathematicians whether they thought about research problems symbolically, or using some kind of mental image. With very few exceptions, they used visual imagery, even when the problem and its solution were mainly symbolic. For example, Hadamard's mental image for Euclid's proof that there are infinitely many primes involved not algebraic formulas, but a confused mass to represent the known primes, and a point far from that mass to represent a new prime. Vague metaphorical images were common, formal diagrams like Euclid's rare.

The tendency to invoke visual (and tactile) images is evident as far back as Al-Khwarizmi's *Algebra*, whose title refers to 'balancing'. The image invoked is one that teachers often use today. The two sides of an equation are thought of as collections of objects placed in the corresponding pans of a pair of scales, which must balance. Algebraic operations are then performed in the same manner on both sides, to ensure that it remains balanced. Eventually we end up with the unknown quantity in one pan and a number in the other: the answer. Mathematicians solving equations often imagine the symbols moving around. (That's why they still like blackboard and chalk: a bit of rubbing out and rewriting can achieve much the same effect.) More obviously geometric thinking also occurs in Al-Khwarizmi's *Algebra*, with its diagram of the process of completing the square to solve a quadratic equation. According to legend, one mathematician gave a very technical lecture on algebraic geometry, drawing just a single dot on the blackboard to represent a 'generic point'. He referred to it frequently, and the lecture made far more sense as a result. Blackboards and whiteboards across the planet, not to mention napkins and sometimes tablecloths, are crammed with a jumble of esoteric symbols and weird little doodles. The doodles can represent anything from a ten-dimensional manifold to an algebraic number field.

Hadamard estimated that about 90 per cent of mathematicians think visually, and 10 per cent think formally. I know of at least one prominent topologist who has trouble visualising three-dimensional shapes. There's no universal 'mathematical mind' – one size doesn't fit all. Most mathematical minds don't proceed one logical step at a time; only the polished proofs of their results do that. Usually, the first step is to get the right idea, often by thinking vaguely about structural issues, leading to some kind of strategic vision; the next is to come up with tactics to implement it; the final step is to rewrite everything in formal terms to present a clean, logical story (Gauss's removal of the scaffolding). In practice most mathematicians alternate between these two ways of thinking, resorting to imagery when it's not clear how to proceed, or when trying to get a simple overview, but resorting to symbolic calculation when they know what to do but are unsure where it leads. Some, however, seem to plough ahead regardless, using only symbols.

Extreme mathematical ability doesn't correlate strongly with anything else. It seems to strike at random. Some, such as Gauss, 'get it' when they're three years old. Some, among them Newton, fritter their childhood away but blossom later in life. Young children generally enjoy numbers, shapes, and patterns, but many lose interest as they grow older. Most of us can be trained in mathematics up to high-school level, but few go beyond. Some never really get to grips with the subject at all. Many professional mathematicians have a strong impression that when it comes to mathematical ability, we're not all born equal. When you've gone through life finding most school mathematics easy and obvious, while others struggle with the basics, it sure looks that way. When some of your students find easy concepts baffling, while others grasp hard ones immediately, this feeling is reinforced.

Perhaps this anecdotal evidence is wrong. A lot of educational psychologists think so. There's been a vogue in psychology for the 'blank slate' view of the child's mind. Anyone can do anything: all they need is training and lots of practice. If you want it badly enough, you'll get it. (If you don't get it, that proves you didn't want it ... a neat piece of circular reasoning greatly favoured by sports commentators.) It would be wonderful if this were true, but Steven Pinker took this kind of politically correct hope apart in *The Blank Slate*. Also, many educators detect a disability, discalculia, which impairs the learning of mathematics like dyslexia affects reading and writing. I'm not sure both positions can be maintained simultaneously.

Physically, we're not all born the same. But for some reason, a lot of people seem to imagine – or want to imagine – that mentally, we are. This makes little sense. Brain structure affects mental abilities just as bodily structure affects physical ones. Some people have eidetic memories that remember everything in great detail. It seems implausible that anyone can be taught to have an eidetic memory if only they train and practise. The blank slate hypothesis is often justified by pointing out that almost everyone who is highly successful in some area of human activity practises a lot. That's true – but it doesn't imply that everyone who does a lot of practice in some area of human activity will be highly successful. As Aristotle and Boole well knew, 'A implies B' is not the same as 'B implies A'.

Before you get too annoyed, I'm not arguing against trying to teach mathematics, or anything else, to anyone. Almost all of us improve with good teaching and plenty of practice, whatever the activity. That's why education is worth the effort. George Pólya revealed some useful tricks in *How to Solve It*. It's a bit like those 'how to have a super-power memory' books, teaching you techniques that help you remember things, but directed towards solving mathematical problems. However, people with eidetic memories don't use mnemonic tricks. What they want to recall is *there* as soon as they need it. Similarly, even if you master Pólya's bag of tricks, you're unlikely to become a new Gauss, however much work you put in. The Gausses of this world don't need to be taught special tricks. They invented them for themselves in the cradle.

On the whole, people don't make themselves successful by working their socks off at something in which they have no real interest. They practise hard because even natural talent needs plenty of exercise to keep it healthy; because you have to keep in practice to stay talented; but mainly because that's what they want to do. Even when it's difficult or boring, in some curious way they enjoy it. You can only stop born mathematicians doing mathematics by locking them up, and even then they'll scratch equations on the walls. And that, ultimately, is the common thread that runs through all of my significant figures. They love their mathematics. They're obsessed with it. *They can do no other.* They give up more profitable professions, they go against their families' advice, they plough on regardless even when many of their own colleagues consider them mad, they're willing to die unrecognised and unrewarded. They lecture for years for no salary, just to get a foot in the door. The significant figures are significant because they're *driven.*

What makes them that way?

It's a mystery.

Notes

1. When referring to a book or paper originally in Latin or a European language, I usually use the English translation of the title, except when historians commonly use an abbreviated form of the original one. The first time such a work appears I give the original title, with a translation, unless it's obvious. The titles of ancient Chinese, Arabic and Indian texts are transliterated and often abbreviated, and a translation is usually provided.

2. George Gheverghese Joseph. *The Crest of the Peacock*, I.B. Tauris 1991.

3. Alexandre Koyré. An unpublished letter of Robert Hooke to Isaac Newton, *Isis* **43** (1952) 312–337.

4. The Royal Society planned to commemorate Isaac Newton's tercentenary in 1942, but World War II intervened so the celebrations were postponed to 1946. Keynes had written a lecture 'Newton, the man', but he died just before the event took place. His brother Geoffrey read the lecture on his behalf.

5. Richard Aldington. *Frederick II of Prussia, Letters of Voltaire and Frederick the Great*, Letter H7434, 25 January 1778, Brentano's 1927.

6. More precisely, the polynomial must also be irreducible – not a product of two polynomials of smaller degree with integer coefficients. If n is prime then $x^{n-1} + x^{n-2} + \ldots + x + 1$ is always irreducible.

7. Boeotia is a region in central Greece. In classical times the Athenians described the Boeotians as dull and unintelligent. The name became a proverbial reference for dumb stupidity.

8. Tony Rothman. Genius and biographers: the fictionalization of Évariste Galois, *American Mathematical Monthly* **89** (1982) 84–106.

9. In UK English the American spelling is often used to distinguish a computer program from any other kind of programme, and is standard in the industry.

10. June Barrow-Green. *Poincaré and the Three Body Problem*, American Mathematical Society, Providence 1997.

11. Ramanujan's Master Formula states that if

$$f(x) = \sum_{k=0}^{\infty} \frac{\varphi(k)}{k!} (-x)^k$$

is a complex-valued function then

$$\int_0^{\infty} x^{s-1} f(x) dx = \Gamma(s) \varphi(-s)$$

where $\Gamma(s)$ is Euler's gamma-function.

12. Andrew Economou, Atsushi Ohazama, Thantrira Porntaveetus, Paul Sharpe, Shigeru Kondo, Albert Basson, Amel Gritli-Linde, Martyn Cobourne, and Jeremy Green. Periodic stripe formation by a Turing mechanism operating at growth zones in the mammalian palate, *Nature Genetics* (2012); DOI: 10.1038/ng.1090.

13. Benoit Mandelbrot. *A Maverick's Apprenticeship, The Wolf Prizes for Physics,* Imperial College Press 2002.

14. See note 13.

15. Benoit Mandelbrot. Information theory and psycholinguistics, in R. C. Oldfield and J.C. Marchall (eds.), Language, Penguin Books 1968.

16. See note 13.

17. Let $c = x + iy$ be a complex number. Start at $z_0 = 0$ and iterate the function $z^2 + c$, getting

$$z_1 = \left(z_0^2 + c\right)$$

$$z_2 = \left(z_1^2 + c\right)$$

$$z_3 = \left(z_2^2 + c\right)$$

and so on. Then c lies in the Mandelbrot set if and only if all the

· points z_n lie in some finite region of the complex plane. That is, the set of iterates is bounded.

18. https://www.youtube.com/watch?v=wO61D9x6lNY.

Further Reading

General Reading

Eric Temple Bell. *Men of Mathematics*, Simon and Schuster 1986. (First published 1937.)

Carl Benjamin Boyer. *A History of Mathematics*, Wiley 1991.

Morris Kline. *Mathematical Thought from Ancient to Modern Times*, Oxford University Press 1972.

MacTutor History of Mathematics archive: http://www-groups.dcs.st-and.ac.uk/~history/

Wikipedia: https://en.wikipedia.org/wiki/Main_Page

Chapter 1   Archimedes

Eduard Jan Dijksterhuis. *Archimedes*, Princeton University Press 1987.

Mary Gow. *Archimedes: Mathematical Genius of the Ancient World*, Enslow 2005.

Thomas L. Heath. *The Works of Archimedes* (reprint), Dover 1897.

Reviel Netz and William Noel. *The Archimedes Codex*, Orion 2007.

Chapter 2   Liu Hui

George Gheverghese Joseph. *The Crest of the Peacock*, I.B. Tauris 1991.

Chapter 3   Muhammad al-Khwarizmi

Ali Abdullah al-Daffa. *The Muslim Contribution to Mathematics*, Croom Helm 1977.

George Gheverghese Joseph. *The Crest of the Peacock*, I.B. Tauris 1991.

Roshdi Rashed. *Al-Khwarizmi: The Beginnings of Algebra*, Saqi Books 2009.

Chapter 4   Madhava of Sangamagrama

George Gheverghese Joseph. *The Crest of the Peacock*, I.B. Tauris 1991.

Chapter 5   Girolamo Cardano

Girolamo Cardano. *The Book of My Life*, NYRB Classics 2002. (First published 1576.)

Girolamo Cardano. *The Rule of Algebra (Ars Magna)* (reprint), Dover 2007. (First published 1545.)

Chapter 6   Pierre de Fermat

Michael Sean Mahone. *The Mathematical Career of Pierre de Fermat, 1601–1665* (second edition), Princeton University Press 1994.

Simon Singh. *Fermat's Last Theorem – The Story of a Riddle that Confounded the World's Greatest Minds for 358 Years* (second edition), Fourth Estate 2002.

Chapter 7   Isaac Newton

Richard S. Westfall. *The Life of Isaac Newton*, Cambridge University Press 1994.

Richard S. Westfall. *Never at Rest*, Cambridge University Press 1980.

Michael White. *Isaac Newton: The Last Sorcerer*, Fourth Estate 1997.

Chapter 8   Leonhard Euler

Ronald S. Calinger. *Leonhard Euler – Mathematical Genius in the Enlightenment*, Princeton University Press 2015.

William Dunham. *Euler – The Master of Us All*, Mathematical Association of America 1999.

Chapter 9   Joseph Fourier

Ivor Grattan-Guinness. *Joseph Fourier 1768–1830*, MIT Press 1972.

John Hervel. *Joseph Fourier – the Man and the Physicist*, Oxford University Press 1975.

Chapter 10   Carl Friedrich Gauss

Walter K. Bühler. *Gauss – A Biographical Study*, Springer 1981.

G. Waldo Dunnington, Jeremy Gray, and Fritz-Egbert Dohse. *Carl Friedrich Gauss: Titan of Science*, Mathematical Association of America 2004.

M.B.W. Tent. *The Prince of Mathematics – Carl Friedrich Gauss*, A.K. Peters / CRC Press 2008.

Chapter 11   Nikolai Ivanovich Lobachevsky
Athanase Papadopoulos (editor). *Nikolai I. Lobachevsky, Pangeometry,* European Mathematical Society 2010.

Chapter 12   Évariste Galois
Laura Toti Rigatelli. *Évariste Galois 1811–1832* (Vita Mathematica), Springer 2013.

Chapter 13   Augusta Ada King
Malcolm Elwin. *Lord Byron's Family: Annabella, Ada and Augusta, 1816–1824,* John Murray 1975.
James Essinger. *Ada's Algorithm – How Lord Byron's Daughter Ada Lovelace Launched the Digital Age,* Gibson Square Books 2013.
Anthony Hyman. *Charles Babbage – Pioneer of the Computer,* Oxford University Press 1984.
Sydney Padua. *The Thrilling Adventures of Lovelace and Babbage – The (Mostly) True Story of the First Computer,* Penguin 2016.

Chapter 14   George Boole
Desmond MacHale. *The Life and Work of George Boole* (second edition), Cork University Press 2014.
Gerry Kennedy. *The Booles and the Hintons: Two Dynasties That Helped Shape the Modern World,* Atrium 2016.
Paul J. Nahin. *The Logician and the Engineer: How George Boole and Claude Shannon Created the Information Age,* Princeton University Press 2012.

Chapter 15   Bernhard Riemann
John Derbyshire. *Prime Obsession: Bernhard Riemann and the Greatest Unsolved Problem in Mathematics,* Plume Books 2004.
Marcus Du Sautoy. *The Music of the Primes: Why an Unsolved Problem in Mathematics Matters* (second edition), HarperPerennial 2004.

Chapter 16   Georg Cantor
Amir D. Aczel. *The Mystery of the Aleph: Mathematics, the Kabbalah, and the Search for Infinity,* Four Walls Eight Windows 2000.
Joseph Warren Dauben. *Georg Cantor: His Mathematics and Philosophy of the Infinite* (second edition), Princeton University Press 1990.

Chapter 17 Sofia Kovalevskaia
Ann Hibner Koblitz. *A Convergence of Lives – Sofia Kovalevskaia: Scientist, Writer, Revolutionary*, Birkhäuser 1983.

Chapter 18 Henri Poincaré
Jean-Marc Ginoux and Christian Gerini. *Henri Poincaré: A Biography Through the Daily Papers*, WSPC 2013.

Jeremy Gray. *Henri Poincaré, A Scientific Biography*, Princeton University Press 2012.

Jacques Hadamard. *The Psychology of Invention in the Mathematical Field*, Princeton University Press 1945. (Reprinted Dover 1954.)

Ferdinand Verhulst. *Henri Poincaré*, Springer 2012.

Chapter 19 David Hilbert
Constance Reid. *Hilbert*, Springer 1970.

Ben Yandell. *The Honors Class: Hilbert's Problems and Their Solvers* (second edition), A.K. Peters / CRC Press 2003.

Chapter 20 Emmy Noether
Auguste Dick. *Emmy Noether: 1882–1935*, Birkhäuser 1981.

M.B.W. Tent. *Emmy Noether: The Mother of Modern Algebra*, A.K. Peters / CRC Press 2008.

Chapter 21 Srinivasa Ramanujan
Bruce C. Berndt and Robert A. Rankin. *Ramanujan: Letters and Commentary*, American Mathematical Society 1995.

Robert Kanigel. *The Man Who Knew Infinity – A Life of the Genius Ramanujan*, Scribner's 1991.

S.R. Ranganathan. *Ramanujan; The Man and the Mathematician* (reprint), Ess Ess Publications 2009.

Chapter 22 Kurt Gödel
Gabriella Crocco and Eva-Maria Engelen. *Kurt Gödel, Philosopher-Scientist*, Publications de l'Université de Provence 2016.

John Dawson. *Logical Dilemmas: The Life and Work of Kurt Gödel*, A.K. Peters / CRC Press 1996.

Chapter 23 Alan Turing
Andrew Hodges. *Alan Turing: The Enigma*, Burnett Books 1983.

Michael Smith. *The Secrets of Station X: How the Bletchley Park Codebreakers Helped Win the War*, Biteback Publishing 2011.

Dermot Turing. *Prof: Alan Turing Decoded*, The History Press 2016.

Chapter 24   Benoit Mandelbrot

Michael Frame and Nathan Cohen (eds.). *Benoit Mandelbrot: a Life in Many Dimensions*, World Scientific, Singapore 2015.

Benoit Mandelbrot. *The Fractalist: Memoir of a Scientific Maverick*, Vintage 2014.

Chapter 25   William Thurston

David Gabai and Steve Kerckhoff (eds.). William P. Thurston, 1946–2012. *Notices of the American Mathematical Society* **62** (2015) 1318–1332; **63** (2016) 31–41.

Index